Praise for

The Scoop on Poop

Our current system of discarding nutrient-rich human waste is shameful (and foolish) and it's high time we learned how we can close this vital piece of Mother Nature's nutrient cycles. As a leading expert on homesteading and solar how-to, Dan Chiras is the perfect person to tackle this important topic.

— Cheryl Long,
Editor in Chief, *Mother Earth News*

With world population doubling rapidly, California wilting into a desert, and the planet warming by degrees, we have to ask — so, you have a large bowl of drinking water in a room in your house, maybe even a few such bowls in a few such rooms. And whenever these bowls are empty, you just push a button, flick a lever, or pull a chain and they automatically refill. And you do what with this water? Say what?

— Albert Bates
author of *The Post-Petroleum Survival Guide and Cookbook (2006)* and
The Biochar Solution: Carbon Farming and Climate Change (2010).

Dan Chiras has covered every topic under the sun, so it should be no surprise that he tackles that last piece of the sustainability loop, and that is the concept of waste. Although "waste" is, as Dan illustrates, not an accurate term, because of all the benefits of capturing and recycling the nutrients in our … output. Dan demystifies the process and addresses the misconceptions to teach the reader how to create systems to treat our waste, which is extremely important to those trying to get "off the grid". I highly recommend this book for anyone ready to close the loop on their poop.

— James R. Plagmann,
Architect + LEED AP, HumaNature Architecture, LLC

In his usual cheerful style, clean energy guru and master ecologist, Dan Chiras, tackles the smelly, but important, subject of closing the loop on our personal waste. He shows us how to take excrement and greywater and create a valuable agricultural resource to be returned to the natural cycle of our gardens. If you want to take your composting regime to a whole new level, you seriously need this book!

— Sylvia Bernstein,
author of *Aquaponic Gardening: A Step-by-Step Guide to Raising Vegetables and Fish Together*

Let Dan Chiras and *The Scoop on Poop* encourage you to take an interest in cleaning up after yourself. This book is a great introduction to taking personal responsibility for waste water issues, and can put you on the road to drastically lowering your own contributions to this large scale problem.

—Chris Magwood, Sustainable building educator,
The Endeavour Centre

the Scoop on Poop

the Scoop on Poop

SAFELY CAPTURING and RECYCLING the NUTRIENTS in GREYWATER, HUMANURE, and URINE

Dan Chiras, Ph.D.

Illustrations by Forrest Chiras

Cover design by Diane McIntosh.
Illustrative elements © iStock
Interior Illustrations by Forrest Chiras. Interior photos by Dan Chiras except where indicated. Sidebar background: © iStock alexdans

Printed in Canada. First printing February 2016.

Funded by the Government of Canada | Financé par le gouvernement du Canada | Canada

Paperback ISBN: 978-0-86571-787-9
eISBN: 978-1-55092-583-8

Inquiries regarding requests to reprint all or part of *The Scoop on Poop* should be addressed to New Society Publishers at the address below. To order directly from the publishers, please call toll-free (North America) 1-800-567-6772, or order online at www.newsociety.com

Any other inquiries can be directed by mail to:
New Society Publishers
P.O. Box 189, Gabriola Island, BC V0R 1X0, Canada
(250) 247-9737

New Society Publishers' mission is to publish books that contribute in fundamental ways to building an ecologically sustainable and just society, and to do so with the least possible impact on the environment, in a manner that models this vision. We are committed to doing this not just through education, but through action. The interior pages of our bound books are printed on Forest Stewardship Council®-registered acid-free paper that is **100% post-consumer recycled** (100% old growth forest-free), processed chlorine-free, and printed with vegetable-based, low-VOC inks, with covers produced using FSC®-registered stock. New Society also works to reduce its carbon footprint, and purchases carbon offsets based on an annual audit to ensure a carbon neutral footprint. For further information, or to browse our full list of books and purchase securely, visit our website at: www.newsociety.com

LIBRARY AND ARCHIVES CANADA CATALOGUING IN PUBLICATION

Chiras, Daniel D., author

The scoop on poop : safely capturing and recycling the nutrients in greywater, humanure and urine / Dan Chiras, Ph.D. ; illustrations by Forrest Chiras.

Includes index.
Issued in print and electronic formats.
ISBN 978-0-86571-787-9 (paperback).--ISBN 978-1-55092-583-8 (ebook)

1. Graywater (Domestic wastewater). 2. Water reuse. 3. Urine. 4. Feces. I. Title.

TD653.C45 2016 363.72'8 C2015-907139-9
C2015-907140-2

Contents

Preface:

Warning!

This book is not for the squeamish!

Years ago, a friend of mine opened a college-level human biology textbook I wrote. As chance would have it, the book opened to the chapter on female reproduction. In fact, the pages that almost always appeared when anyone cracked open the book were those with graphic anatomical illustrations of "lady parts" (scientifically referred to as external genitalia), legs spread wide apart as if for a gynecological exam for everyone and their cousin to see. This happened time and time again, so much so that it seemed a bit like a plot by some naughty and playful cosmic force.

Upon his second or third "flashing" and his subsequent examination of the two chapters on human reproduction, my friend commented, "This book seems to have a lot on sex."

Before I could respond, however, he realized how silly his statement was. I can't recall his exact words, but it was something like, "Of course, that's appropriate. Reproduction is a vital function and a key part of human biology."

He was right, of course. None of us would be here were it not for reproduction. It's just that so many people are uneasy with sex — especially when reminded that we're the result of an act performed by dear old mom and dad. Now, that's downright yucky!

The Scoop on Poop is another book that dwells on an often embarrassing part of human physiology — excretion. That said, I offer a stern warning: This book is not for the squeamish or faint of heart, as it is chock full of references to human feces and urine and the processes that produce these icky products. However, if you are concerned about the fate of humankind and are interested in learning how to live sustainably and self-sufficiently on planet Earth, I urge you to look past this apparent effrontery.

The Scoop on Poop tackles a subject few of us think about: the value of feces, urine, and greywater. Distilled to its basics, this book is about consciously altering our lives to stop squandering the valuable nutrient-rich excretions we produce each and every day of our lives, and safely returning them to soil where they can nourish plants that feed us. This idea may be gross to many people, but it is really quite natural. The process mimics Mother Nature's "waste" recycling system and offers enormous benefits to humankind.

If the memory of a smelly campground latrine or a portable potty on a hot summer day at an outdoor concert immediately pops into your mind, capturing, composting, and reintegrating human "wastes" into soil may seem primitive and repulsive. Or, if you've ever been on a long western river trip where you were required to carry all of your group's poo out in a "honey bucket," you're probably gagging at the thought of recycling humanure and urine and applying the composted materials to your garden.

Most residents of more developed countries like Canada and the United States have grown accustomed to the flush toilet. A quick flick of the handle washes all that yucky stuff down the drain, and we're done with it.

To the skeptical, let me say that this book is not about stockpiling or handling raw human sewage. It's about designing systems that safely and odorlessly compost our "wastes," converting them to sanitary soil amendments. This book will teach you how to reclaim and utilize your bodily "wastes" and how to make this process as painless, odorless, and "unyucky" as possible.

If you have ever changed a baby's diapers, emptied a kitty litter box, or scooped dog doo up in a plastic bag from a city park or a neighbor's lawn, you'll be pleasantly surprised at how much easier and more pleasant it can be to compost human excrement — if you do it right!

But, once again, I warn you: this book is not for squeamish or queasy prissy-folk who gag at the slightest malodor. If you are one of these people, put this book down immediately. Pass it on to a friend with a stronger stomach, or return it to the bookstore for a refund.

If you decide to read on, please do so with an open mind. Don't assume the worst. My goal is to introduce you to techniques that are clean, painless, and odorless. You won't be lugging around smelly buckets of excrement and urine. Nor will you need to don a hazmat suit and gas mask to successfully compost and recycle your "wastes." Remember, too: as unsavory as this activity may seem, it is one of the most important ecological and sustainable actions you can take to live in harmony with nature — that is, to fit better within the cycles of nature that ensure the continuation of all life on planet Earth.

If you want to pursue a sustainable lifestyle, you can't reach that goal simply by installing a solar electric system, growing organic vegetables, insulating your home, driving an electric car, and recycling cans, bottles, and paper. True sustainability can only be achieved by ensuring that your excretions make it back into their rightful place in nature — back in the soil that nourishes all life on planet Earth — yours included.

Many of you are already taking some of the actions I just mentioned. If you find the notion of dealing with waste to be too challenging right now, perhaps you can put this book and the instructions it contains aside for a while. You may come around to it later. But for those of you who are ready to become even better stewards of the planet and create a more ecologically sound, sustainable, and self-sufficient lifestyle, let's get to it.

Mother Nature Gets an A, We Get an F

1

As soon as there is life, there's waste — lots of it.

It's true of all species, from the simplest single-celled amoeba to the majestic grizzly bear: all living things ingest food and excrete wastes.

Firmly perched at the top of the food chain, humans are no exception to the rule. It doesn't matter whether you are born into an affluent family in a wealthy North American country or to a less-fortunate family in a developing country like Haiti. Collectively, the more than 7 billion human beings that call planet Earth their home produce lots of waste — mountains of waste. The wealthier the nation, the more prolific their waste production.

All told, humans excrete an estimated 1.12 trillion pounds of feces per year. That's 560 million tons a year. To put this otherwise useless statistic into perspective, daily fecal production would fill the Superdome in New Orleans nearly eight times a day — every day of the year. That's a lot of shit!

Urine production is an astounding 3.1 trillion liters per year. Daily urine production from the world's people would fill 3,400 Olympic-sized swimming pools. Over a year's time, urine production would fill nearly 1.25 *million* Olympic-sized pools. It's amazing that we are not up to our knees in pee!

Clearly, we humans produce a great deal of "personal waste."

Humankind has dealt with its "waste" with varying degrees of success since our emergence as a species. Managing our waste, however, became an enormous challenge once we ceased our

nomadic ways and took up residence in cities and towns — a trend that started in the early years of the Agricultural Revolution approximately 10,000 years ago. As human settlements grew, so did the quantities of smelly waste.

Today, there's special urgency to the matter of human bodily excretions. With over 7.3 billion world residents in 2016, and more and more of us packed in crowded cities, human excrement has become a nightmare. With the global human population expanding by about 80 million people per year, it's safe to say that what we do about our excrement and urine could play a huge role in determining our future. The massive outpouring of feces and urine, combined with other global challenges such as exponential growth of the human population, climate change, species extinction, and soil erosion, could all combine to create the demise of human society.

The steps we take to address "waste" and other critical issues will determine whether we build a secure and prosperous future or fall flat on our face and slide down the same slippery dead-end road the dinosaurs traveled 65 million years ago. Lest we forget, we humans come with the exact same warranty as our reptilian predecessors. While dinosaurs were subjected to external forces beyond their control, we humans are actually creating conditions that could lead us to extinction.

In our efforts to create a sustainable path, there's an ecological lesson we need to constantly bear in mind. That is, in nature, there is no such thing as "waste." In an ecosystem, what we think of as "wastes" are valuable nutrient-rich food sources for a large number of organisms. Succinctly put, in nature, all "waste" is food.

"Waste" provides nutrients to decomposer organisms such as bacteria, worms, and insects. Their excretions, in turn, help return nutrients from the "wastes" of living organisms to the soil and water. These nutrients are vital to plants and algae.

Plants and algae, in turn, form the base of all food chains on planet Earth. Put another way, they feed virtually all living organisms. Food chains provide an avenue for nutrients and energy to flow from plants and algae to the rest of the living world. Food chains nourished by nutrients incorporated by plants and algae bind us all together in the web of life — that make our lives possible. As a result, referring to the excretions of organisms as "waste" misses a key ecological principle, and it does this valuable resource a monstrous disservice. In nature's frugal economy, nothing goes to waste, so using the term "waste" is not just inappropriate, it's just plain wrong.

To call excretions of organisms "waste," misses a key ecological principle, and it does this valuable resource a monstrous disservice.

Over the course of biological evolution, Mother Nature has "devised" a plethora of strategies for survival. "Waste" recycling is one of her crowning achievements. Intricate local, regional, and global systems ensure the recycling of essential nutrients, among them carbon, nitrogen, water, and dozens of other chemicals vital to the survival and reproduction of all life forms. Excretions are recycled over and over so life can continue. Life can only continue if the nutrients in bodily excretions are placed back into the system and back into service. It's that simple (Figure 1-1).

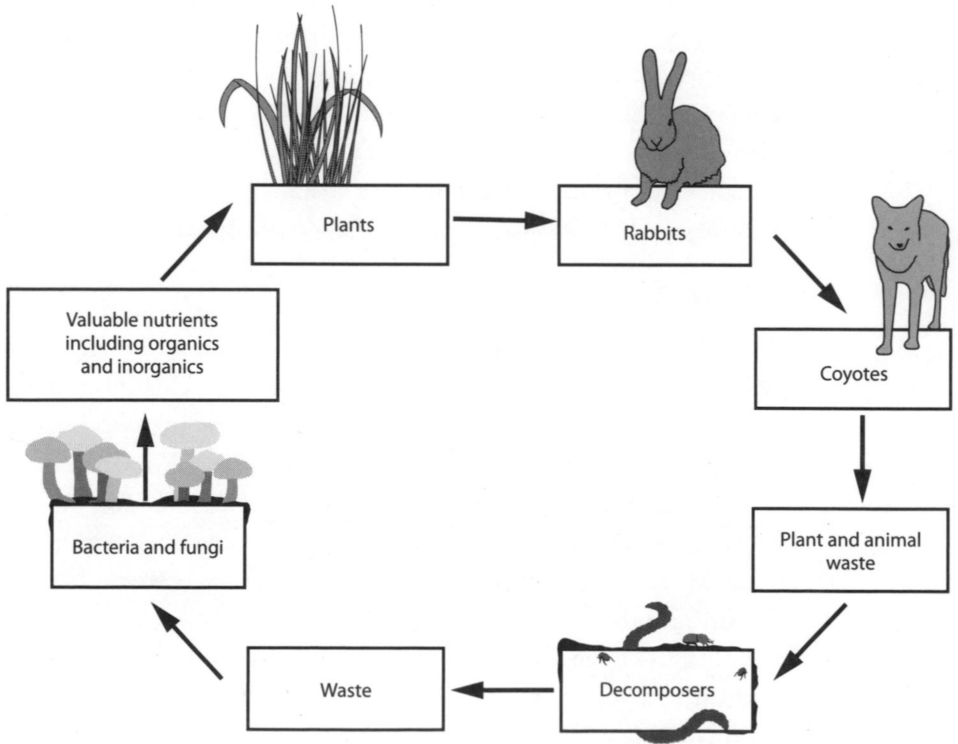

Fig. 1-1: In Nature, There Is No Waste. *All nutrients travel from the environment through organisms in the grazer and decomposer food chains. In the latter, they are returned to the soil for reuse, a process that ensures life's continuation.*

As I remind my ecology students, all life is built on the dead remains of the past. Every molecule in your body was once part of another organism — very likely hundreds of others. Those tears in your eyes may once have been in the blood that flowed through the veins of a prehistoric fish or turtle.

It's for this reason that in this book I don't refer to urine and feces as "waste" without quotation marks, signaling to you what "waste" really is — nutrient-rich material we must recycle to ensure the continuation of life on planet Earth. The only time that calling human excretions "waste" is appropriate, as you shall soon see, is when it refers to the fact that we waste so much of it.

It's a shame that, in our rush to create human settlements, few societies recognized the value of human excretions produced each and every day and that humans failed to design an ecologically intelligent system that would allow us to put our nutrient-rich "waste" to good use, as Mother Nature intended.

The next section, "A Tale of Two Toilets," illustrates two diametrically opposed strategies by which humans have dealt with "waste." One strategy continues today in many parts of the world and undermines our long-term future. The other, from our ancient past, is a model that could, if adopted in large scale, greatly enhance our prospects for building a sustainable future. You get to choose which strategy you'd like to be a part of.

A Tale of Two Toilets

In the late 1800s, the city of Chicago was — quite literally — a cesspool. Teeming with immigrants struggling to eke out a living in a crowded and dangerous city, and hordes of hogs and cattle that arrived by train to end their lives in slaughterhouses, this fetid conglomeration of people and doomed livestock excreted mountains of feces and rivers of urine that either poured directly into Lake Michigan or were dumped into the Chicago River. It then flowed into the lake.

Chicago's large Union Stockyard "processed" 12 million hogs and cattle a year in the 1890s and consumed 500,000 gallons of fresh water per day from the Chicago River. Urine and feces from the stockyards drained into a south fork of the river. It was dubbed "Bubbly Creek" by residents because of the gas bubbles created by the tons of rotting manure that contaminated the waterway.

That was the state of "waste" disposal in Chicago and dozens of other cities throughout the world. Rivers served as sewers. It's not much different from today, though we've cleaned up the effluent we pour into our rivers — at least in many of the more developed countries of the world.

In Chicago, two miles offshore, in the murky depths of the lake, were the city's water intake cribs. Concerned that the flow of human and animal excrement into the lake was too close to the city's water source and that pathogens in the waste might lead to an outbreak of cholera or other deadly infectious disease, and under intense public pressure, well-meaning but ecologically ignorant officials ordered the construction of a canal to divert smelly stockyard and human "waste" away from Lake Michigan. Known as the Chicago Sanitary and Ship Canal, it diverted water from the south branch of the Chicago River to the Des Plaines and Illinois Rivers. From these rivers, the wastes flowed southward into the Mississippi River and on to the Gulf of Mexico (Figure 1-2).

Fig. 1-2: The Great Sewage Diversion. *In the drawing on the left, you can see where sewage and waste from livestock flowed into Lake Michigan, the city's source of drinking water. The Chicago Sanitary and Ship Canal, labeled "Canal," in the figure on the right, diverted that water to the Mississippi River, a move like sweeping dust under a carpet.*

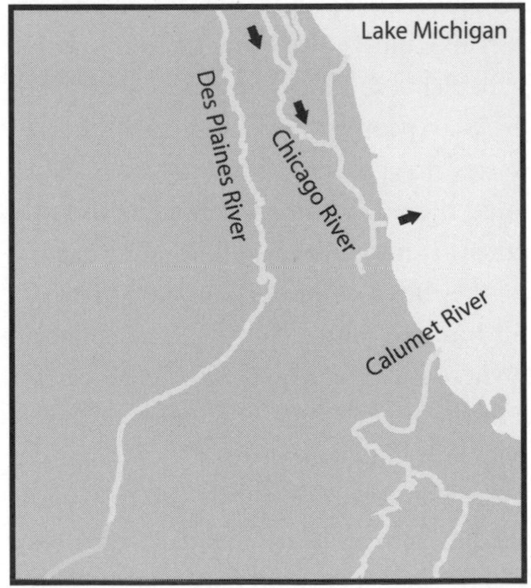

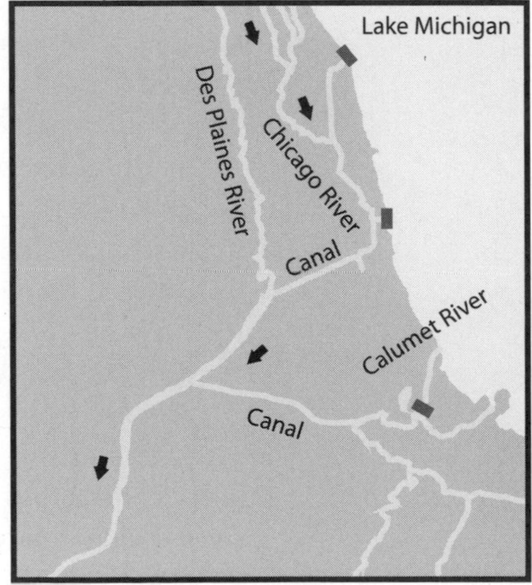

With a single measure, city officials solved an enormous and potentially lethal problem.

Brilliant, eh?

Well, almost.

In typical human fashion, this measure ignored an important detail. Their stroke of genius simply diverted their toxic problem to the mighty Mississippi River. This great waterway became the unlucky recipient of the city's massive daily outpouring of feces and urine from humans and livestock operations. The diversion canal was akin to sweeping dust under a rug. Making matters worse, the steady flow of "waste" from Chicago was augmented along its course by raw sewage released from the numerous rapidly growing cities and towns that dotted the river's banks.

Across the globe, in China, human "waste" management was fashioned in a distinctly more ecological manner. Here, rural peasants carried nutrient-rich excrement in buckets from their homes to the land on which they planted crops that fed their families. These nutrients nourished the soil and ensured a steady, sustainable supply of food. The practice continues today.

So, in rural China, excrement was viewed for what it was: a valuable agricultural resource. The Chinese were not the only people who recycled the valuable nutrients contained in their "waste." For centuries, numerous other Asian peoples have religiously returned nutrients from human "wastes" to agricultural soil. In Japan, prior to World War II, for instance, human "waste" was bought and sold as fertilizer. In fact, excrement from wealthy individuals purportedly sold for more because they ate a more nutrient-rich diet.

For centuries, nearly all human "waste" in Asia was collected and delivered to the soil where it decomposed. In the process, it added a wide assortment of extremely valuable organic and inorganic nutrients to life-giving farm and garden soil.

But don't think that Asia had a corner on the human waste recycling market. In England, human and livestock waste and household garbage (such as kitchen scraps) were regularly hauled out of the city and returned to farm soil. This job fell to a group of independent contractors, known as *night-soil men.* The system was designed, in part, to reduce the putrid stench produced by a city of over 1.4 million people. In 1815, cities had no other means of "waste" disposal.

Mike Rowe (the host of the Discovery Channel's "Somebody's Gotta Do It") would probably agree that those night-soil men had the dirtiest jobs ever devised. The crews had to retrieve excrement from public and private cesspits. To access the nutrient-rich material, one member of the four-man crew, the holeman, descended into the pit by ladder. He scooped up liquid sludge from the pit, and poured it into larger buckets. The buckets full of feces and urine were then lifted to the surface by another worker, the ropeman. He transferred the "waste" into a larger tub that was loaded on a cart and hauled out of the city. The "waste" was dumped on farmland. This practice went on well into the mid-1800s, when the city began installing sewer lines.

If Mother Nature were to grade the two resource management plans, Chicago would have received an F; Asia an A+.

As any organic gardener knows, organic matter like that contained in human excrement builds soil. One reason for this is that organic matter in soil acts like a sponge — that is, it helps to retain moisture, helping plants to prosper.

Organic matter also creates a nutrient-rich environment that supports a healthy population of beneficial soil microorganisms — bacteria and single-celled fungi, for instance. Some of these organisms convert the ammonia in animal waste to nitrite. Nitrite is then converted by other bacterium into nitrate. It can be taken up by plants and used to produce nitrogen-containing compounds like nucleic acids that form the genetic material of plants and amino acids that are used to build proteins.

A host of other nutrients, like phosphorus and calcium, are also released from urine and feces. These nutrients are taken up by plant roots. Plants use these nutrients to grow, reproduce, and produce fruit and seeds. As pointed out earlier, plants form the base of grazer food chains, from which all animals — including human beings — receive nutrition. Also noted earlier, plants feed most of the biological world. Urine and feces from animals are a nutritional staple of virtually all organisms on the planet.

As many of us learn, but few fully appreciate, in nature, animal excretions are a nutritional staple of virtually all organisms.

Effluenza — Poisoning Our World, Poisoning Ourselves

Despite the exalting human accomplishments — art, architecture, music, and technology — that suggest to many that humans are the crowning achievement of nature, human beings are a dangerous force. We're not just making a mess of the planet, we are poisoning

The Scoop on Poop

Many people are surprised to find out what's in their feces. To begin with, human feces are largely made of water (75%). If you dry feces out, its weight declines by 75%.

The remaining 25% are solids. Of the solids, about 30% consists of intestinal bacteria — a horde of benign but highly useful bacteria that dine on undigested foodstuffs. These bacteria, for the most part, provide a wide range of valuable services to their human hosts.

Another 30% consists of undigested food matter. That's food that either can't be digested, such as cellulose (the water-insoluble fiber in celery, for instance) or food that makes it through the digestive tract without being digested — that is, broken down by enzymes.

Feces also contain about 10% fat, including cholesterol. A small amount of our feces — about 2% to 3% — is protein. Feces also contain dead cells shed from the inner lining of the intestinal tract.

What makes up the remainder?

The remaining 10 to 20% of human feces consists of valuable inorganic substances such as calcium phosphate and iron phosphate.

Why are feces brown? Feces' characteristic brown color derives from a chemical called *bilirubin*. It's nothing more than a breakdown product of the blood protein, hemoglobin. Bilirubin comes from aged red blood cells the body retires each day. As you may recall from high school biology, hemoglobin is the protein in red blood cells to which oxygen binds. Bacteria in the intestine chemically modify hemoglobin, turning it to brown bilirubin.

What about the odor given off by feces? Odor is caused by a handful of chemicals that are released as intestinal bacterial do their job of digesting food in feces.

ourselves and the millions of species that share this planet with us. We are rapidly depleting the Earth's resources, treating the Earth as if it were a corporation in liquidation. Leading the liquidation sale are the most developed of all nations. The most affluent have become the most *effluent.*

Modern resource-intensive societies are foreclosing on the human future by directing the steady stream of perfectly usable materials — like millions upon millions of tons of municipal solid waste (trash) containing valuable biological and technological nutrients from aluminum and steel to carbon and nitrogen — into landfills, which are more or less permanent tombs. Air pollution, water pollution, sewage, mine "wastes," and factory-generated toxic chemical "wastes" add to our continuous planetary assault. But in this book, I'll focus on the "wastes" that issue from two of humankind's most vital excretory systems, the gastrointestinal tract and the urinary system.

On average, an adult human being produces between a quarter and a half pound of feces every day, depending on one's sex, food consumption, weight, and other factors. (For those using metric system, that's about 100 to 250 grams per day.) Over a year's time, we excrete 90 to 180 pounds of shit a year. Translated: most of us produce about our body weight in crap every year. (For a discussion of the contents of feces, see the accompanying textbox.)

Straight Line Thinking in a Circular World

Few people who have studied the current plight of humankind in depth would disagree: we humans have charted an unsustainable course. We're depleting renewable resources such as trees, fish, and soil faster than they can be replaced. We're literally eating and consuming ourselves out of house and home. And, we're producing numerous types of pollution at rates that far exceed the planet's ability to dilute and detoxify them. As a result, we're poisoning ourselves and the species that share this planet with us. Some of our pollutants are also altering conditions like temperature and climate on planet Earth that are vital for survival. In the process, we humans are taking the remainder of the living world down with us.

The roots of our unsustainability are many. First and foremost, we've built a massive, complex, resource-intensive civilization that depends on nature but doesn't seem to recognize or respect this vital dependence. Put another way, we've built a society that gets what it needs from the Earth and ecosystems but does far too little to maintain the health of the ecosystems we depend on, which are quite literally the life-support systems of the planet.

Another reason we are on an unsustainable course is that human society continually violates the laws of nature. Chief among our foibles is a reliance on *linear thinking.*

Over many years, linear thinking has shaped human civilization. Linear thinking has spawned a multitude of unsustainable linear systems that include the human economy, agriculture, waters supply, manufacturing, and "waste management" — to name a few.

Linear systems are an anathema in nature. They fail miserably in a cyclic world.

To understand the fundamental flaw of linear thinking and linear systems design, consider an example: manufacturing. To provide the goods we need, such as paper or building materials, corporations extract resources from the environment (Figure 1-3). They cut down trees, then process the wood into forms that enable us to manufacture a wide assortment of goods such as writing paper and framing lumber. These goods are sold and put to use. All is well and good ... so far.

Unfortunately, when their useful life is over — or the products we devour have fallen out of fashion — we dump them in landfills. Landfills are fancy names for dumps that serve as tombs

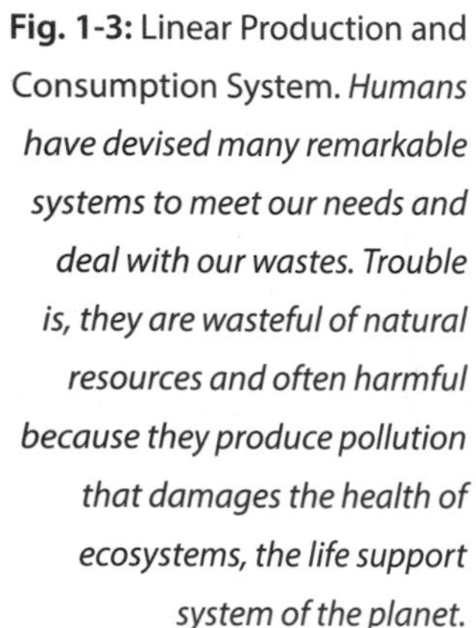

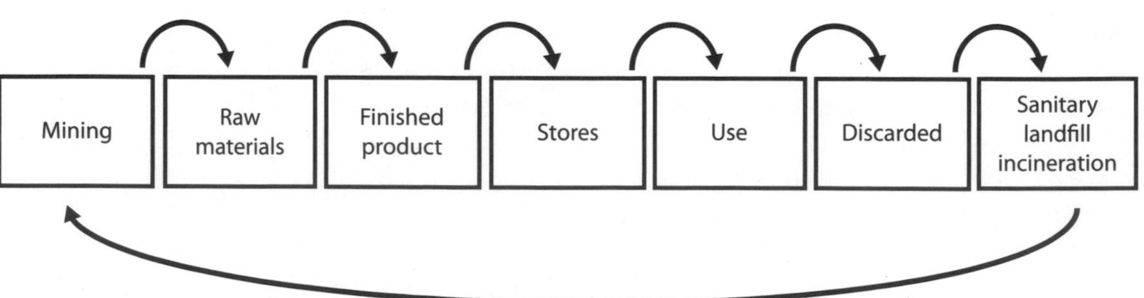

Fig. 1-3: Linear Production and Consumption System. *Humans have devised many remarkable systems to meet our needs and deal with our wastes. Trouble is, they are wasteful of natural resources and often harmful because they produce pollution that damages the health of ecosystems, the life support system of the planet.*

where our "wastes" remain locked up pretty much for life — unless we choose to mine them in the future (Figure 1-3).

No matter whether it is a cell phone, a television, a pair of jeans, a plastic beach ball, or a lawn chair, all products follow the same unecological path. Resources flow from the Earth's crust and from natural and human-altered ecosystems (like farms and forests) to manufacturing facilities. From there they move to distribution outlets, stores, and then on to us, the eager consumers. When the products fall apart or out of favor, it's off to their earthen grave. The system of production and consumption has become a one-way path to oblivion.

Global food production throughout much of the world suffers from the same short-sighted, ecologically ignorant thinking. Farmers grow food in topsoil. Manufacturers process the fruits of the earth into edible delights, sometimes removing lots of the nutrients (Figure 1-4). Food is then shipped to our tables via trucks, boats, and trains to distributors to stores to our shopping carts to our cars and our tables. We consume foods, utilizing some of the nutrients in them, like organic foodstuffs (starches) to produce energy molecules like glucose. What we don't utilize, along with products of cellular metabolism, are excreted as feces or urine. We then deposit our "wastes" in a porcelain bowl filled with 3 to 3.5 gallons of clean drinking water. The "wastes" are flushed down the drain and make their way to a local

Fig. 1-4: Linear Food Cycle. *Unlike in nature, our food and the nutrients it contains flows from natural and human-altered ecosystems in an unsustainable linear fashion.*

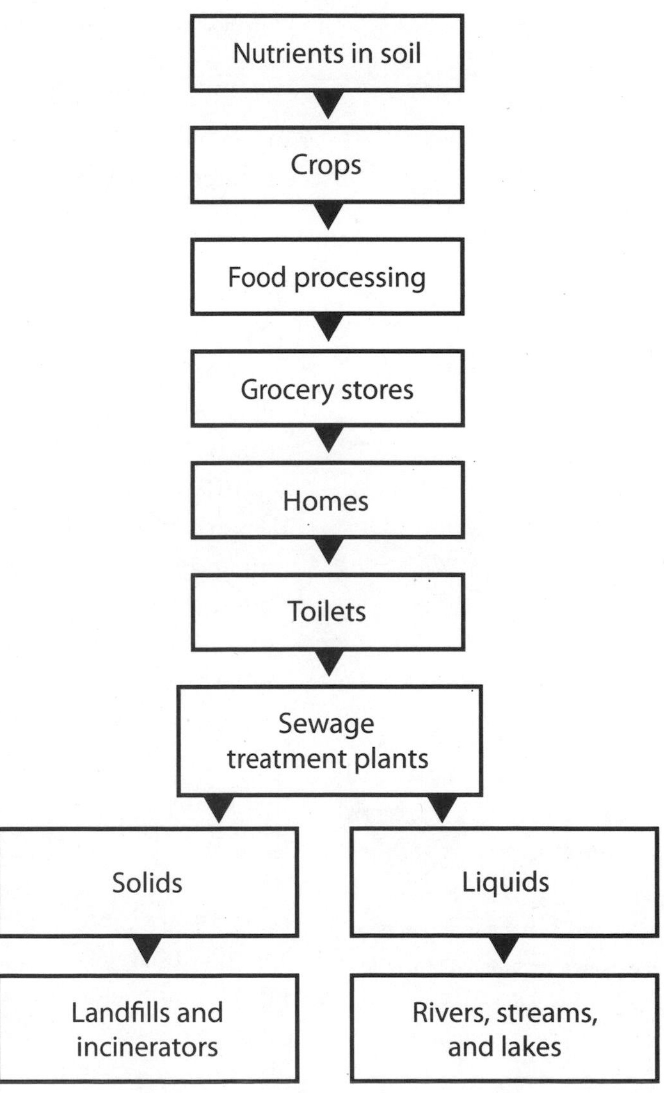

sewage treatment plant. In these facilities, the organic solids are precipitated out, dried, and typically dumped in landfills, alongside diapers, old TV sets, and back issues of *The National Enquirer* (Figure 1-3).

The liquid portion of the "waste" stream passes through additional treatment stages designed to remove most of the pollutants — aka valuable nutrients, such as nitrogen. What's left — the effluent — is then treated with chlorine or other chemicals to kill potential pathogens and dumped into nearby water bodies — lakes, rivers, or oceans (Figure 1-3).

From an engineering standpoint, this system is brilliant. From an ecological standpoint, it is stupid and, worse yet, suicidal. As I pointed out earlier, what is egregiously wrong about this system is that very few nutrients that travel from the soil to plants to humans to sewage treatment return to their site of origin — the farmers' fields.

While some readers may understand the folly of nonlinear systems, most citizens are clueless that our system of food production and consumption — so vital to our survival — is a one-way nutrient pathway to extinction. It slowly but surely depletes the soils of nutrients and dumps many of them into surface waters where in high concentration they can cause disease and upset ecosystems and poison aquatic life or landfills where they can leach into groundwater.

It's easy to reduce this to its bare bones: the "modern" human food production-consumption system is a system of nutrient depletion and pollution. Is the greatest misallocation of natural resources on the planet. And through these linear systems humankind is sowing the seeds of its own destruction.

It's easy to reduce this to its bare bones: the "modern" human food production-consumption system is a system of nutrient depletion and toxic pollution. Is the greatest misallocation of natural resources on the planet. And through these linear systems humankind is sowing the seeds of its own destruction.

In rural areas in the developed world, the picture's much the same — just on a more individual level. Most homeowners flush their "wastes" into 500- to 1,000-gallon septic tanks. Made of concrete, fiberglass, or plastic, septic tanks are typically buried in our backyards close to our homes. Within these now-ubiquitous septic tanks, solids settle to the bottom, slowly forming a thick layer of sludge (Figure 1-5).

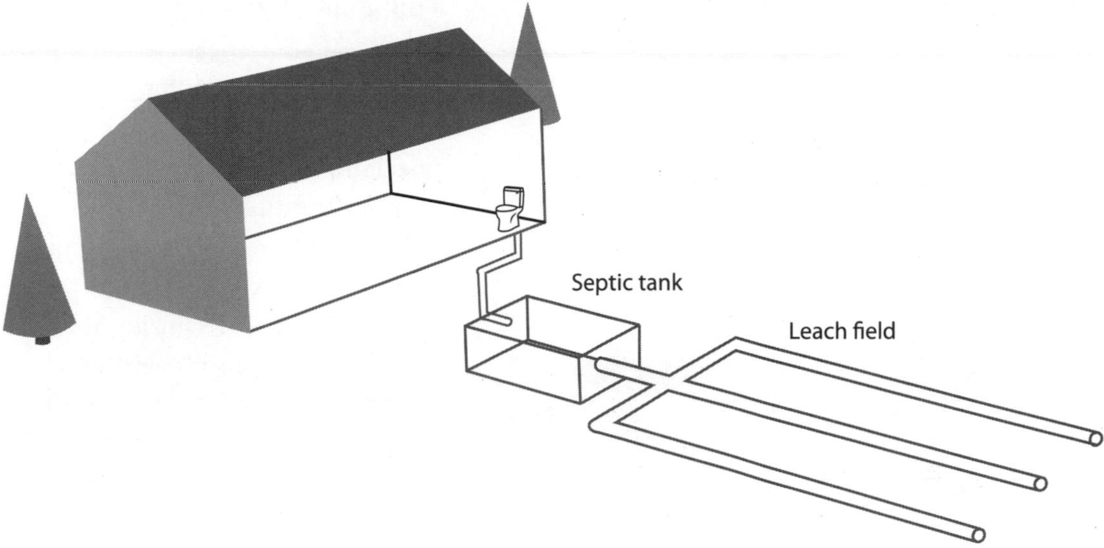

Fig. 1-5: Septic Tank and Leach Field. *Excretions of rural residents in more developed countries are typically deposited in septic tanks. Liquid effluent from the system drains through the leach field into the ground too deeply to be put to good use.*

Liquids containing soaps, grease, dissolved organic material, nutrients, and scum fill the top layer of septic tanks. It flows out of tanks through pipes that lead to leach fields. Leach fields are nothing more than a set of porous pipes buried deep in the ground to disperse liquid "wastes." They are designed to dispose of liquid wastes so deep that they can't rise to the surface where they could cause problems — or nourish plants. In fact, leach fields are designed to drain downward, and hence they become a potential source of groundwater pollution.

The solid "waste" that accumulates in septic tanks is periodically removed by a special pump truck. This prevents the organic matter from overflowing and draining into the leach field, clogging up the works. Organic buildup in the pipes of the leach fields reduces the effectiveness of a field, and can clog the pipes entirely, putting a leach field out of commission.

Linear waste mis-management of material resources, especially food, is one of the key reasons human society is on a one-way road to oblivion.

Organic sludge removed from septic tanks is typically trucked to sewage treatment plants where it is added to municipal sewage coming from homes. It suffers the same fate as urban and suburban sewage. Solids end up in landfills. Liquids end up in surface waterways. Nutrients rarely make it back to soils from which they came.

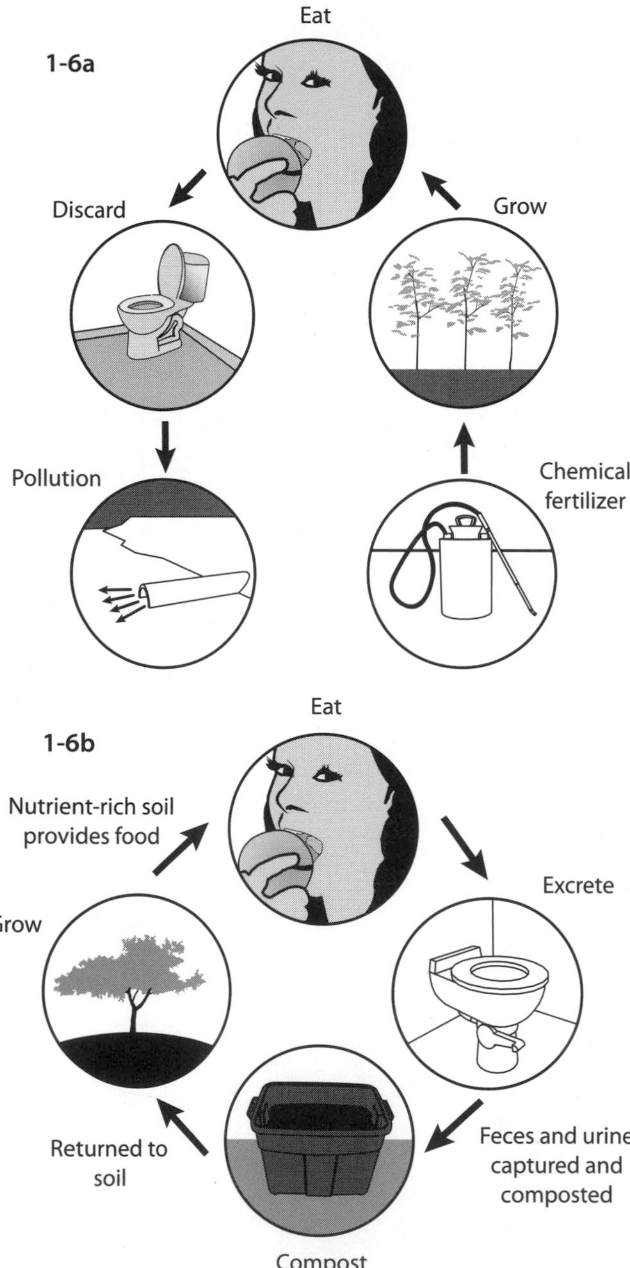

Fig. 1-6a and b: Linear and Cyclic Food Production Consumption. *It doesn't take much for us to individually change the flow of nutrients to ensure sustainable food production.*

Building a Sustainable Society

If we are going to survive and prosper, we need to end linear thinking in *all* aspects of our lives, but especially with respect to food and human bodily excretions (Figure 1-6a). Translated, we need to put our nutrient-rich solid and liquid "waste" to good use — fertilizing the land and building the soils we depend on. In returning nutrients to their site of origin, we can ensure a steady supply of food to feed us and our offspring — and theirs (Figure 1-6b).

This is where you come in.

You can play a role in closing the loops, ensuring cyclicity, in food production and consumption (Figure 1-6b). You can help by recycling your nutrient-rich "waste" carefully and safely reinserting the nutrients back into the topsoil that you rely on to raise the food you and your loved ones eat. No matter whether you live in the suburbs or a city — if you have access to some land where you can compost your waste and grow your own food — you can become a participant in this vital effort.

As Gene Logsdon notes in his book, *Holy Shit,* "We can't truck all our waste back into the country, but we can truck it into our backyard gardens and orchards and create a sustainable lifestyle no matter where we live."

If you live in the country and have plenty of room for these activities, all the better. You can capture your solid and liquid "waste," treat it to be sure it is safe, and work it into the soil. More on this throughout the book.

If you raise animals, even chickens, you'll have even more nutrient-rich manure to revitalize the soil. Raising a cow can produce about 15 tons of manure-soiled bedding each year. Calves will produce about half that. That gives you a lot of valuable material to compost, use, or even sell.

While I have no grand illusions of modern society figuring out that it has to be smarter with its waste and taking actions to recycle these valuable nutrients, I can imagine hundreds of thousands of readers like you taking matters into their own hands ... or buckets, I suppose. I can envision this dedicated legion working to eliminate the concept of "waste" from their lives, building a path to sustainability that others will then emulate.

This book is about helping you find ways to join me and others in recycling the valuable nutrients you're currently wasting. It's about viewing human excretions differently, much differently, not as a vile material to get rid of, but as a resource to put to good use — a material that nourishes the land, animals, and your family.

I'll show you many options that work, give you information to get started, and discuss the pros and cons of each option. I'll tell you how do this safely. I'll discuss my experiences, notably what's worked and what hasn't. I'll share my favorite ideas, my successes, and my failures in my continual quest to return these valuable nutrients to their rightful place.

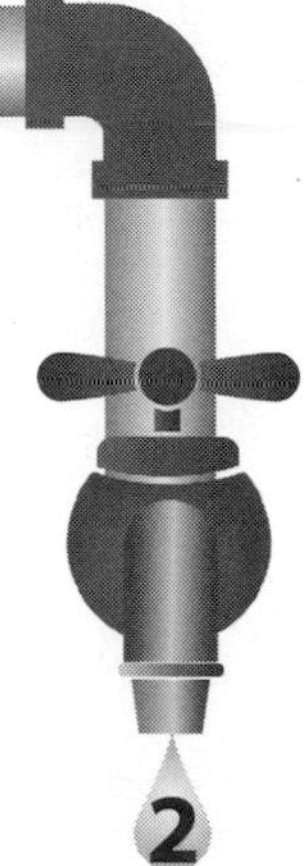

Buying a Composting Toilet

The first requirement for successful return of nutrients in human urine and feces is a clean, reliable, and easy-to-empty system of capture. Over the years, several companies have stepped forward to meet this demand. They've also designed composting toilet systems that will compost the "wastes." Their toilets tend to be sleek and attractive — they look great in most conventional bathrooms (Figure 2-1). Although commercially available composting toilets require more user participation than flush toilets, they've been designed to minimize the amount of work required by the homeowner.

Fig. 2-1: Commercial Composting Toilet. *Commercial composting toilets like this one from Envirolet are attractive and, if maintained properly, fairly easy to use.* Source: Sancor Industries.

In this chapter, I'll examine the more popular commercially available composting toilets — a few of which I've installed and used over the years — and I'll discuss their pros and cons. This information will help you understand your options and what you're getting into — specifically what is required to install and use the system so it performs well. Before you lay your money down, though, I strongly urge you to check out online reviews of each product and talk to individuals who have used these systems. You can learn a lot from other people's experiences. If possible, try out various systems. All this can help you decide which system to buy and how to successfully operate it. A piece of seemingly insignificant information can make the difference between success and failure.

Before we delve into the details of commercially available composting toilets, however, let me define what is meant by the term *composting toilet.*

What's a Composting Toilet?

Most readers who have gardened maintain a compost pile or two, often housing them in compost bins to keep critters out. A compost pile is a natural bioreactor. Gardeners "feed" the pile a variety of plant material, such as leaves, grass clippings, weeds from the garden, and fruit and vegetable peels. Even old blue jeans, scraps of paper, and shredded cardboard can be composted.

In a compost pile, organic materials such as these rapidly decompose under optimum conditions thanks to the action of insects and an army of naturally occurring oxygen-loving (aerobic) bacteria and single-celled fungi. Worms may migrate into the pile to dine on your organic "wastes." As they feed on organic matter, these decomposing organisms excrete a nutrient-rich stream of feces, politely referred to as *worm castings.* Worm castings enrich compost.

The final product of compost piles is referred to as compost. It consists of decayed and decaying organic matter. It makes an excellent soil supplement.

Composting toilets are designed to capture "wastes" and provide an environment that allows naturally occurring microbes to turn human excrement into a fluffy organic material called humus. Humus is nutrient-rich compost that can be used to fertilize soils in our orchards, berry patches, and gardens.

In most composting toilets, feces and urine are deposited in a chamber where they are mixed with additional organic matter such as sawdust or wood shavings. (I dump paper towels and paper napkins in mine, too.) Properly designed and operated, composting toilets convert this mixture of organic materials into humus (not to be confused with hummus, a deliciously creamy blend of chickpeas, oil, garlic, lemon juice, and tahini that makes a great, healthy sandwich, especially when other vegetables like lettuce and tomatoes are added). Humus from composting toilets, on the other hand, is brown or nearly black organic material containing decaying or decayed feces, toilet paper, and sawdust or some other organic additive. It's not as delectable as its near homonym — unless you are a decomposer organism such as a bacterium, single-celled fungi, earthworm, or fruit fly. Then, the stuff is chocolate-covered cherries!

Humus provides an excellent nutrient-rich environment for valuable soil-dwelling nutrient-cycling microbes — microorganisms that are vital to the health of farm and garden soils. When the soil is healthy, it creates an excellent environment for plants that grow in it. As an added bonus, humus also increases the ability of soils to absorb and retain water, which leads to healthier plants. Plants are less likely to be stressed during drought or in dry periods when soils contain more moisture.

Features of Successful Composting Toilets

Well-made composting toilets perform several closely related functions. First and foremost, they provide an odorless, insect-free chamber to collect human excrement. In most models, users deposit urine and feces in the same receptacle. As you shall soon see, some composting toilets are equipped with urine diverters — devices that separate feces and urine. According to many sources, separating urine from feces helps reduce odors.

Composting toilets can also serve as a repository for kitchen scraps. This is especially valuable during cold winter months when the outdoor compost pile is covered in snow or frozen solid. We add soiled kitty litter to our composting toilets in the winter. (We use pine pellets for litter. Be sure not to throw noncompostable clay or artificial litter into a composting toilet. They'll fill it up quickly.)

Secondly, a well-designed composting toilet ensures that moisture in feces and urine readily evaporates. Without adequate ventilation and drying, you'll end up with a soupy liquid. Not only is this liquidy sludge smelly, it's difficult to safely handle.

In most well-designed composting toilets, water evaporates through a vent pipe that exits at roof level so odors won't be detected. Evaporated liquid (water vapor) and odors may be vented either passively or actively, that is, forced out by a small electric fan.

Thirdly, a composting toilet should be designed to create an environment that promotes rapid decomposition of feces, toilet paper, and the organic cover material you'll add each day. This ensures the production of a rich, organic humus in a relatively short period.

Fourth, and of great importance, well-designed and properly operated composting toilets create an environment that destroys potentially harmful viruses, bacteria, fungi, and intestinal parasites. If properly processed, the organic end-product can be handled with little, if any, risk to one's health. Humus from composting toilets can be dug into garden soil, providing valuable nutrients, but I recommend adding composted humanure to a conventional garden compost pile for further processing. This ensures additional decomposition, destruction of pathogenic organisms, and hence greater safety. That's the route we take at home and at my educational center, The Evergreen Institute. (More on this at the end of the chapter.)

Why Compost Humanure?

Composting toilets provide many benefits to people and the environment. Because composting toilets allow us to take control of our own nutrient-rich excretions, they help us significantly reduce the amount of sewage that we produce — "waste" that typically ends up in septic tanks and conventional sewage treatment plants. In so doing, they

---

Benefits of Composting Humanure

Composting toilets are vital to those of us who understand the importance of closed-loop nutrient cycles to the survival of life on planet Earth. They allow us to recycle valuable nutrients in our excretions and put them back where they belong, in the soil, where Mother Nature intended them to go — so they can be put to good use growing food.

---

also decrease our consumption of fresh water required for flushing conventional toilets. Composting toilets also help us reduce the amount of energy required to pump and purify water and process raw sewage. They also assist in reducing groundwater and surface water pollution. Composting humanure helps us build soil and grow lots of healthy food. It helps us become more self-sufficient.

Types of Composting Toilets

Now that you understand what composting toilets do, let's take closer look to enhance your understanding of your options. This will also help you understand how the commercially available composting toilets work.

Composting toilets can be categorized by several criteria. Let's first examine a functional classification.

Active and Passive Composting Toilets

When shopping for a composting toilet, you'll soon discover that they come in two basic varieties: passive or active.

Passive composting toilet systems are much like low-maintenance backyard compost piles. You simply add organic matter — your excretions and an organic cover material such as sawdust, wood chips, ground coconut shells (coconut coir), or dried leaves. These additives boost the carbon content of the humanure. They also create air spaces. Oxygen in the air in these cavities helps aerobic microbes digest the organic materials deposited in composting toilets. These organisms liberate valuable nutrients in our excretions, making them available to plants.

Passive composting toilet systems are so named because they require no fans or heaters to get the job done. Moisture evaporates and escapes out the vent pipe naturally.

Active composting toilet systems, on the other hand, typically employ one or two fans and a heater to accelerate water evaporation and microbial decomposition. Fans blow air over and sometimes through humanure to accelerate the rate of evaporation of water in feces and urine. Fans also help whisk odors out the vent pipe, reducing the chances of odors seeping into your home. (Note,

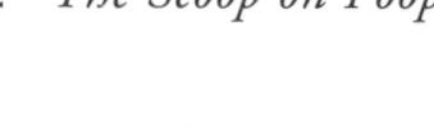

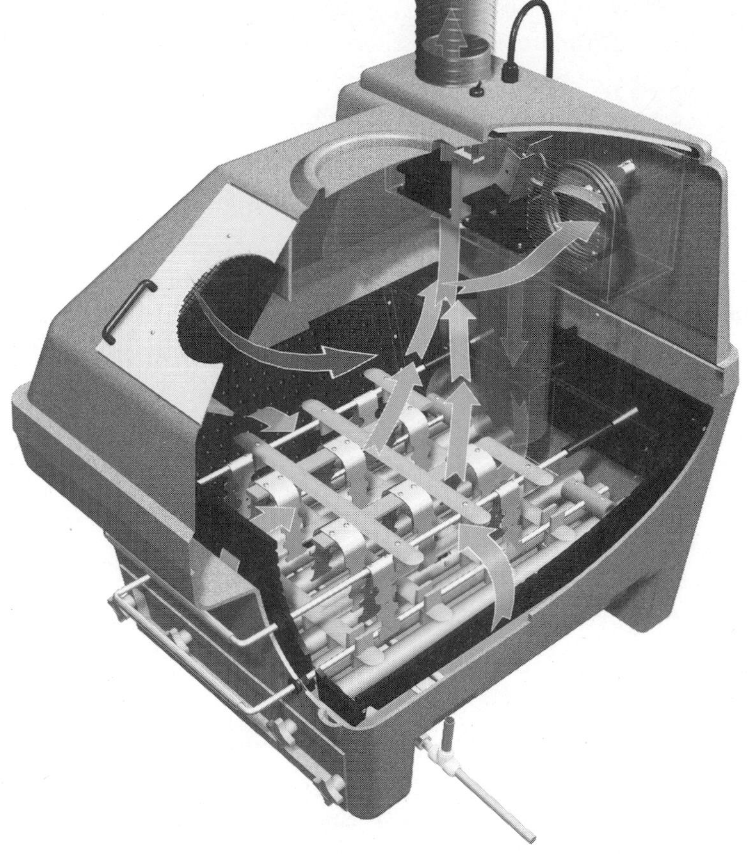

Fig. 2-2: Active Composting Systems. *Fans and heaters as shown in this drawing of an Envirolet composting toilet help remove moisture from feces and urine. Heaters also help promote faster decomposition of organic wastes.* Source: Sancor Industries.

though, that I rarely have problems with odors in my "fanless" systems.) In addition, fans can help "pump" oxygen-rich air into the composting materials, which can facilitate decomposition.

Thermostatically controlled heaters help drive naturally occurring moisture out of the feces and urine deposited in the composting chamber (Figure 2-2). More important, heat helps to maintain optimal temperature inside the composting toilet. That is, it helps maintain a temperature that's conducive to rapid microbial decomposition.

Because they heat and aerate organic matter in their composting chambers, active composting toilets often outperform passive systems. In other words, they typically result in a more rapid and more complete decomposition of the organic materials. Active

Active Systems in Energy-efficient Homes Can Decrease Efficiency

If you are building an airtight, super-efficient home or trying to retrofit a home for energy efficiency, you may want to consider installing a passive composting toilet. Although decomposition of humanure may take longer, passive systems offer many significant energy advantages. They require no outside energy, and they will very likely improve the energy efficiency and comfort of a home. To understand these benefits, let's examine active systems more carefully.

Although fans and heaters accelerate decomposition, they consume electrical energy. In the systems I've installed for use in our home and at The Evergreen Institute's classroom building, I have found that it's not the fans that guzzle the most electricity, it's the heating elements.

In active systems, heaters typically consume between 400 to 600 watts, depending on the manufacturer's choice of heater. Although the heaters are thermostatically controlled and don't usually run 24 hours a day, they can operate a significant amount of time and can significantly increase household electrical energy demand. Increased electrical consumption is especially problematic in homes powered by renewable energy, notably solar electric and wind energy systems.

Consider an example: Envirolet's MS10 composting toilet comes with two 40-watt fans and a 500-watt thermostatically controlled heater. According to the company's website, the unit will operate in the 40-watt mode 75%–80% of the time and at 540 watts 20%–25% of the time. (These estimates depend on the temperature of the room in which the composting chambers are installed.)

Assuming that the toilet operates 18 hours a day at 40 watts and 6 hours a day at 540 watts, a composting toilet will consume about 4 kilowatt hours (kWh) per day — 3.95 kWh to be exact. That's 1,450 kWh a year (365 days per year x 4 kWh per day). In Kansas City, a 1-kW system will provide about 1,400 kWh per year on average — provided it is properly oriented and unshaded throughout the year. A 1-kW grid-tied solar electric system will cost approximately $3,500 to $4,500, depending on the installer and difficulty of installation (at this writing, in May 2015) minus local, state, or federal incentives, if any. In sunnier areas, such as Colorado and New Mexico, the composting toilet fan and heater will require a slightly smaller 0.9-kW system, costing a little under $3,000 to $4,000. If your home is off grid, you'll need batteries and other equipment, easily doubling the system cost. (Remember, ☛

these are the solar electric systems required to run just the composting toilet fans and heater!)

Clearly, composting toilet heaters and fans are pretty pricey features for solar-powered homes, especially for off-grid solar electric homes. They're no bargain if you buy electricity from your local utility, either. If your local utility charges 12 cents per kilowatt-hour, a heater and fan run 10 hours per day will cost you 48 cents a day, $14.40 per month, and $173 per year! If your composting toilet is located in a cold basement, costs and carbon dioxide emissions from coal-fired power plants will be higher. (If you live in a warmer climate, these cost estimates would be lower.)

Another significant downside of active systems with heaters and fans is that they continuously suck air out of our homes. ☛

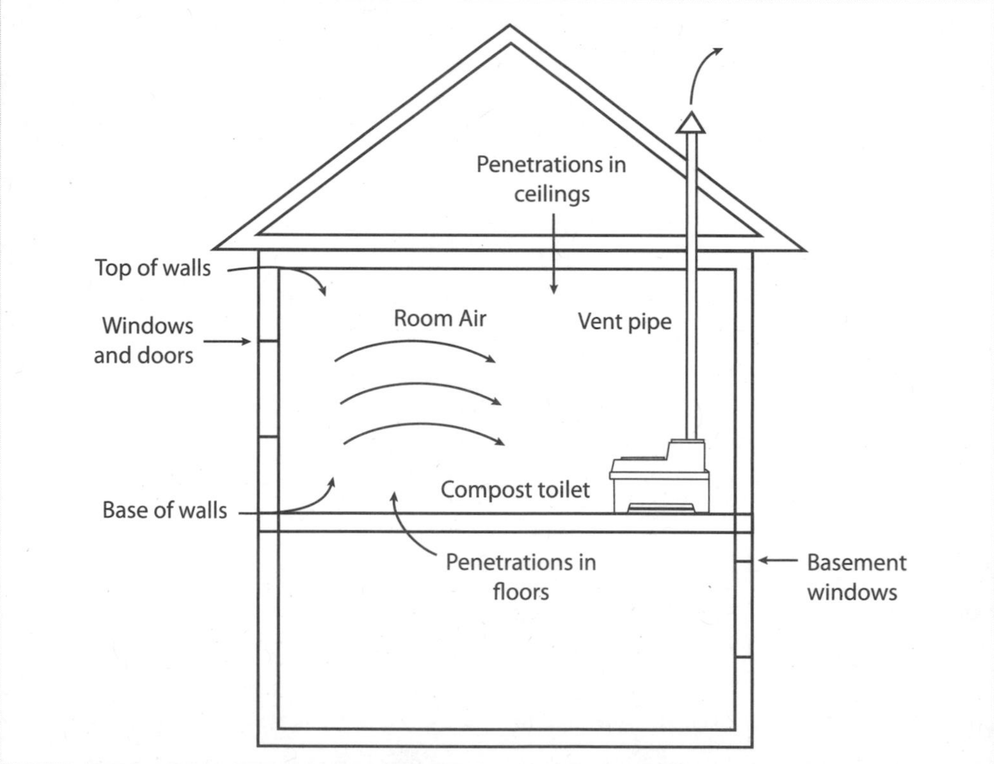

Fig. 2-3: Active Systems and Energy Efficiency. *Active composting toilets remove indoor air that must be replaced by cold outdoor air in the winter and warm outdoor air in the summer, increasing heating and cooling costs.*

When a fan is running, indoor air enters the composting chamber. It runs over and sometimes through the decomposing feces and cover material. This air is then is forced out of the house through the vent pipe. Here's the problem: as show in Figure 2–3, air that is drawn into a composting toilet and expelled via the vent system is replaced by cold outdoor air in the winter. As a composting toilet vents, outdoor air is sucked into a home through many openings in the building envelope — that is, through various cracks and holes in the walls, floors, foundation, and roof. Here's why this is an alarming problem: the average home in the United States has so many leaks in the building envelope that if they were combined they'd be equal to a 3-foot by 3-foot window open 24 hours a day, 365 days a year!

That's right. Your eyes did not deceive you. It's like having a 3-foot by 3-foot window open all the time!

In the winter, warm indoor air is vented to the outside through a composting toilet, and cold outside air is drawn into our sieve-like homes. This cold air makes it more difficult to stay comfortable and greatly increases our demand for energy to heat our homes. It also increases our heating bills and our carbon footprint. My advice is, if at all possible, install a passive system.

systems are especially useful in colder climates — that is, in areas where temperatures routinely or seasonally drop below 55°F (12.8°C). Below this temperature, natural composting grinds to a halt.

Batch or Continuous Systems

Both active and passive composting toilets may be batch or continuous composters. A batch composting toilet is one that includes two or more receptacles to capture urine and feces. When the first one is filled, it is either removed or rotated out of the "line of fire" so the second receptacle can be filled. The first one is allowed to undergo further decomposition without the addition of any more "waste" while the second container fills.

BioLet's 30 NE (nonelectric) composting toilet contains a large internal compost bin. When the bin is approximately three-fourths full, the top of the toilet is removed and the bin is removed.

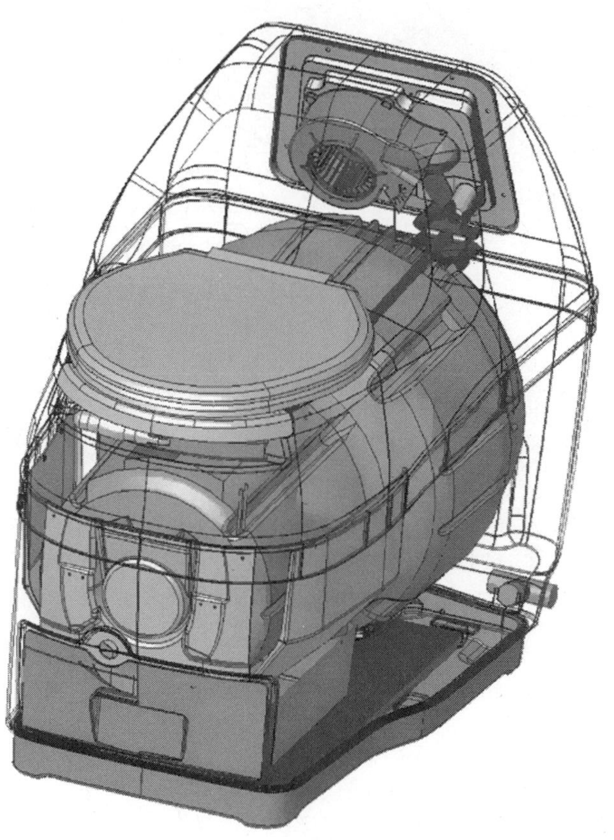

Fig. 2-4: Sun-Mar Composting Toilet. *The rotating drum in this system houses organic bulking agent and feces. Rotation ensures aeration. When the drum needs emptying, it is simply rotated in reverse. This dumps the composted "waste" into a tray for removal.*

It is replaced by a second bin. Meanwhile, the first bin is usually stored outside or in a warm location to permit further composting.

A continuous composter, the more common design, typically contains a single composting chamber. New material is added at the top. As a result, the pile of feces and cover material grow with each use. As new material is deposited, however, organic matter deposited in previous days decays. The continual decay of organic "wastes" slows the pile growth of fecal matter and cover material.

Continuous composting systems are prevented from overflowing by periodically removing composted humanure, typically from the bottom of the composting chamber. This is achieved by one of several ingenious methods. In Sun-Mar composting toilets, the feces and cover material drop into a drum (Figure 2-4). It is periodically rotated to mix feces and the cover material, such as sawdust. Mixing also permits oxygen to enter the composting mass, theoretically accelerating decomposition. When time comes to empty the drum, it is rotated in reverse. This opens a door in the drum, letting the composted material drop into a tray beneath it. The tray can then be removed and emptied.

Envirolet's composting toilets fill from the top and are emptied from the bottom. Humus is liberated from the bottom of the pile by blades attached to two rods that run from the front of the unit to the back (Figure 2-5). They are attached to a handle in the front of the unit. Moving the handle in and out rakes the blades through the humus, causing it to fall into a removable tray,

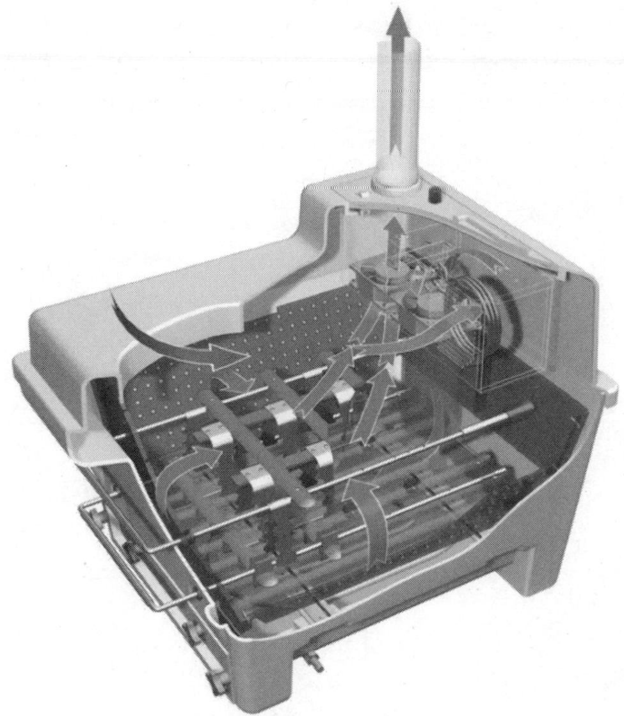

Fig. 2-5: Interior View of Envirolet Composting Toilet, Showing Heaters. *Notice the blades attached to handles are designed to aerate the compost (top) and cause humus to drop into a tray. (below)* Source: Sancor Industries.

accessed through a watertight door. Details on their operation are presented later.

BioLet's electric composting toilets (20 Deluxe and 60 XL models) contain an electric mixer that automatically mixes the feces and cover material, facilitating aerobic decomposition. The mixing arms are controlled by a switch that's activated when one opens the toilet's lid. They run for about one minute after the lid is closed. BioLet's nonelectric model known as "10 Standard" contains a manually operated T-Handle located on the top of the unit for mixing. As usage continues, compost accumulates inside the compost chamber. The composted humanure drops into a tray below through a trap door-type device (damper) controlled by a manually operated lever.

Self-contained vs Remote Composting Toilets

So far, you've learned that composting toilets may be active or passive and batch or continuous. Composting toilets may also be

Fig. 2-6: Self-contained Composting Toilet. *In this system, the toilet seat sits on top of the composting chamber.*

self-contained or remote. Each of these options can be either active or passive and batch or continuous. Let's start with self-contained composting toilets.

As shown in Figure 2-6, a self-contained composting toilet consists of a throne (a place for you to sit), a toilet seat, a composting chamber, and tray/drawer in which humus is collected and removed. As a rule, self-contained composting toilets are rather large and bulky. They can become a 900-pound gorilla in a tiny bathroom. Some folks find them too out of the ordinary (odd-looking) to install in a conventional bathroom. Users sit up high on a plastic throne above the composting chamber. It's not exactly a regal throne worthy of your nobility, but it is odorless, sanitary, and much more ecologically sound than conventional flush toilets. To me, the self-contained composting toilet is a sign of distinction — a sign of deep caring and action. It's the sacred place where one's ethics manifest themselves in action.

Remote composting toilet designs place the receptacle in separate room, usually a basement below the throne (Figure 2-7). They are ideal for homes with basements, and work well in new construction where you can plan the layout so bathrooms

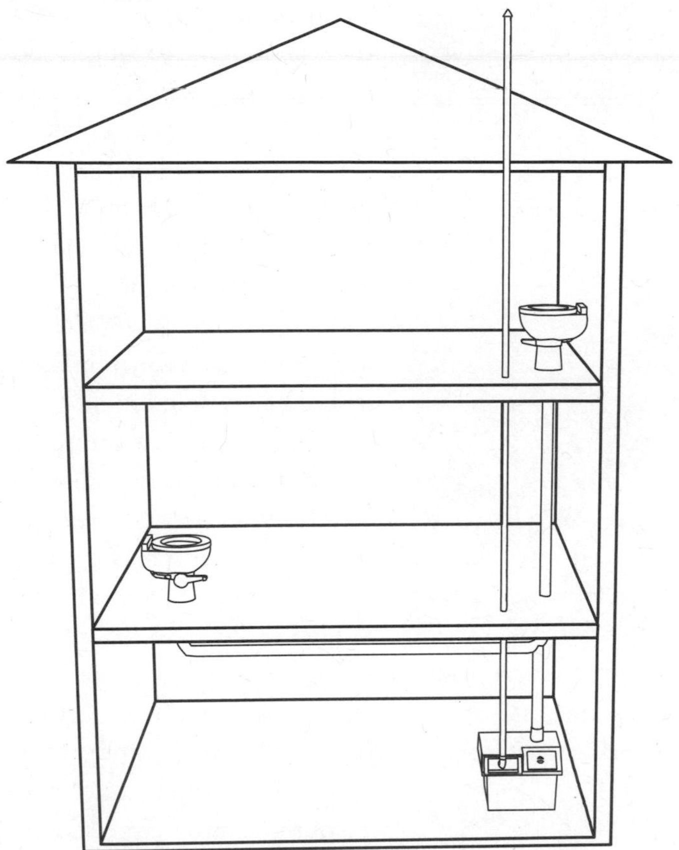

Fig. 2-7: Remote Composting Toilet System. *Remote systems allow more than one compost toilet to be used. These systems may be wet or dry. They often include one-pint flush toilets, as in the system shown here.*

are located over strategically placed composting chambers in the basement.

The toilet in remote composting systems consists of small throne, a seat, and a lid. I find many models, especially those produced by Envirolet, to be quite attractive, as you can see from the photo of my bathroom (Figure 2-8).

Remote composting toilets may be either dry or wet. In dry composting toilet designs, feces and urine are deposited in the toilet and fall directly into the composting chamber located directly below. This is the type of system we installed in our home in 2013.

Most homeowners who install dry, remote composting toilets place the composting chamber in their basements directly below the throne. In a two-story home without a basement, the throne

Fig. 2-8: Remote Composting Toilet Throne. *The throne for the remote waterless composting system is sleek and attractive. It goes well with most bathroom decors.*

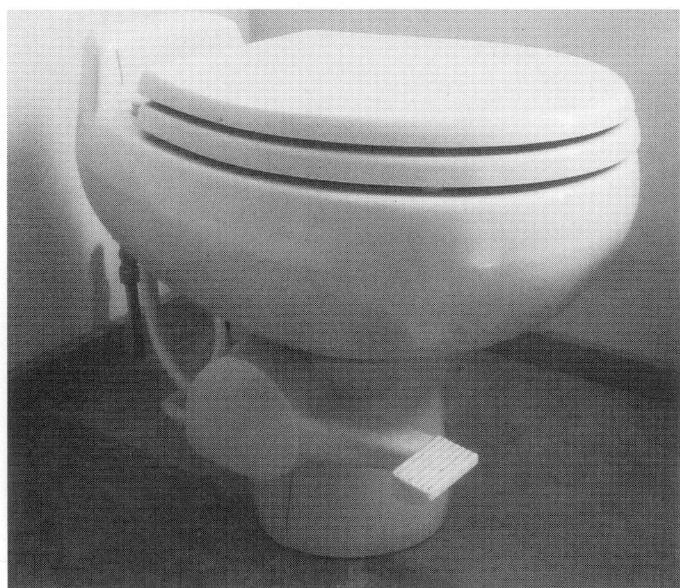

Fig. 2-9: One-Pint Wet Remote Composting Toilet.

could be located in a bathroom on the second story, so long as the composting receptacle is placed directly below it. That's rarely possible.

Wet remote composting toilets are designed to provide more flexibility in placing toilets and composting chambers. These units are equipped with an extremely low-water-use toilet like the one shown in Figure 2-9 in TEI's classroom building. Because urine and feces are transported in water, the composting chamber can be located in a location not directly below the toilet. This design is especially helpful in homes with multiple toilets, as shown in Figure 2-7.

As you can see in Figure 2-9, water-efficient, low-flush toilets look pretty nice. What is more, they only require one pint of water per flush. To put this into perspective, a modern water-efficient flush toilet uses 1.28 to 1.6 gallons — nearly 10 to 13 pints of water! Some older toilets use 3 to 3.5 gallons per flush — or 24 to 28 pints per flush!

At TEI, we leave the toilet bowl empty between uses. When time comes to make a deposit, students simply lift up on the foot-operated lever located on the side of the unit. This allows water to flow into the bowl. When a pint has entered, the foot lever is released. Students do their thing, then press down on the foot pedal to flush. Urine and feces with a small amount of water flow by pipes into the composting chamber below. A fan and heater help evaporate water; however, a drain pipe is almost always required to remove excess liquid from the composting chamber in such systems.

Most readers have been raised on flush toilets. Because they are so used to using large quantities of water to flush feces and urine down the tubes, many are skeptical about the efficacy of one pint of water to flush. It seemed too good to be true to me, too, but after using the remote wet composting system in our classroom building for several years, I've found that this tiny amount of water does the job beautifully. Students seem to like them, as they operate much like conventional toilets. (Like dogs, people in more developed countries seem to be mighty particular about how and where they defecate!)

The one-pint flush toilets in TEI's bathrooms were purchased from the Canadian company, Sun-Mar. We also purchased the main receptacle/composter from them. Sun-Mar's one-pint flush toilets are well designed and attractive. Very rarely are there any "skid marks" on the bowl from errant fecal matter. These toilets are easy to use and easy to clean. We are continually astounded at how little water they use. Because of this, they're ideal for off-grid homes whose water comes from rain catchment systems. They also save energy in homes equipped with wells, because lower water demand means less energy required to pump water.

The only problem we've had with our system is annoying overflows — sewage seeping out on the basement floor. This often happens when I host a workshop with a half dozen or more students. We've found that if there's a kink in the drain hose, liquid containing flush water, dissolved poo, and urine leaks out of the composting chamber and onto the floor of the basement. (You'd think the manufacturer would design them to be more watertight!)

My advice is this: if you are considering a remote system with one-pint flush toilets, size it appropriately — so it can handle the extra usage on special occasions when large numbers of people show up for a visit. Also be sure to install a reliable, bomb-proof, permanent drain system (don't rely on a hose that runs out through a partially opened door like we do). Put a drain system in right away to handle heavy use and prevent overflow that may occur when you entertain large groups, for example, at Thanksgiving.

Outdoor Composting Facilities

Most composting toilets are designed for in-home installation and use. They're ideal for cabins, cottages, workshops, and homes — if sized properly. If you'd like, you can also build an outhouse composting toilet or install a commercially available composting toilet in an outdoor facility.

Don't get confused. An outdoor or outhouse composting toilet is a whole different animal than a traditional outhouse — aka latrine (Figure 2-10). Latrines of yesteryear are nothing more than an upright wood structure perched over a hole in the ground. Latrines contained a throne or two for one or two users.

In latrines, users deposit urine and feces into the pit until it's full. At that time, a new pit is dug and the house is repositioned. The excrement and urine-filled pit is covered with dirt and left to rot, providing little, if any, benefit to the environment or the owner.

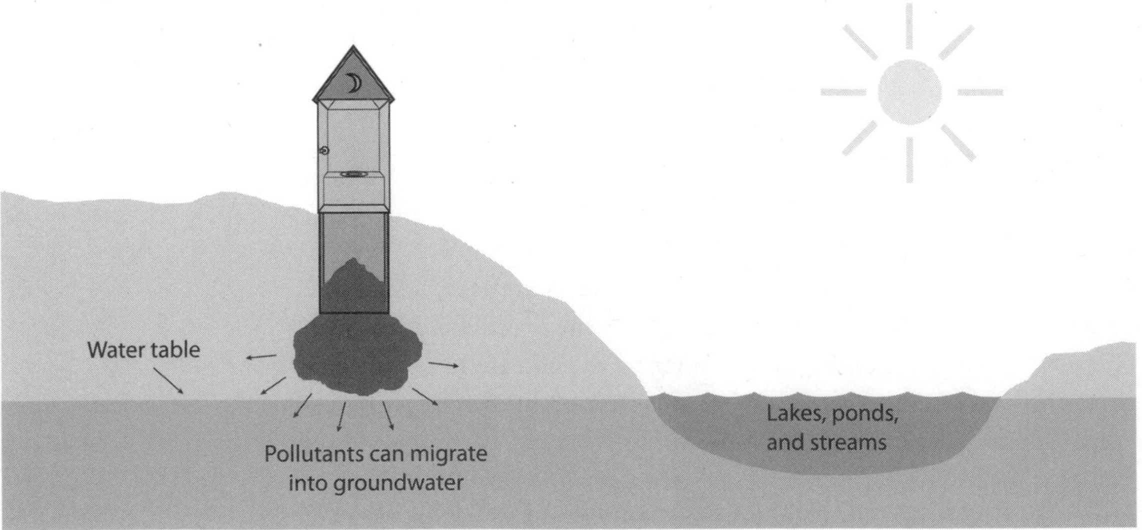

Fig. 2-10: Latrines are Notorious Polluters. *Latrines are nothing more than outhouses built over open pits into which people deposit their feces and urine. They are notorious sources of superficial groundwater pollution.*

Latrines are notoriously smelly and renowned for polluting shallow groundwater in their immediate vicinity. In dry soils, polluted water can migrate as far as 10 feet (3 meters) downward.

Outdoor composting toilets, on the other hand, capture nutrient-rich excretions from users in a watertight receptacle (Figure 2-11). It is mixed with organic cover material and toilet paper, and the composted material is periodically emptied. As a result, there's little, if any, risk of contaminating the surrounding soil and groundwater.

Outdoor composting toilets are best used in warmer climates where temperatures inside the composting chamber remain above 55°F (12.8°C) year round for optimal function. That said, it's ok if compost freezes in the winter. In warm weather, it will thaw and continue aerobic digestion.

Another option for an outdoor composting toilet is the Clivus Multrum (CM), shown in Figure 2-12. Since its invention in

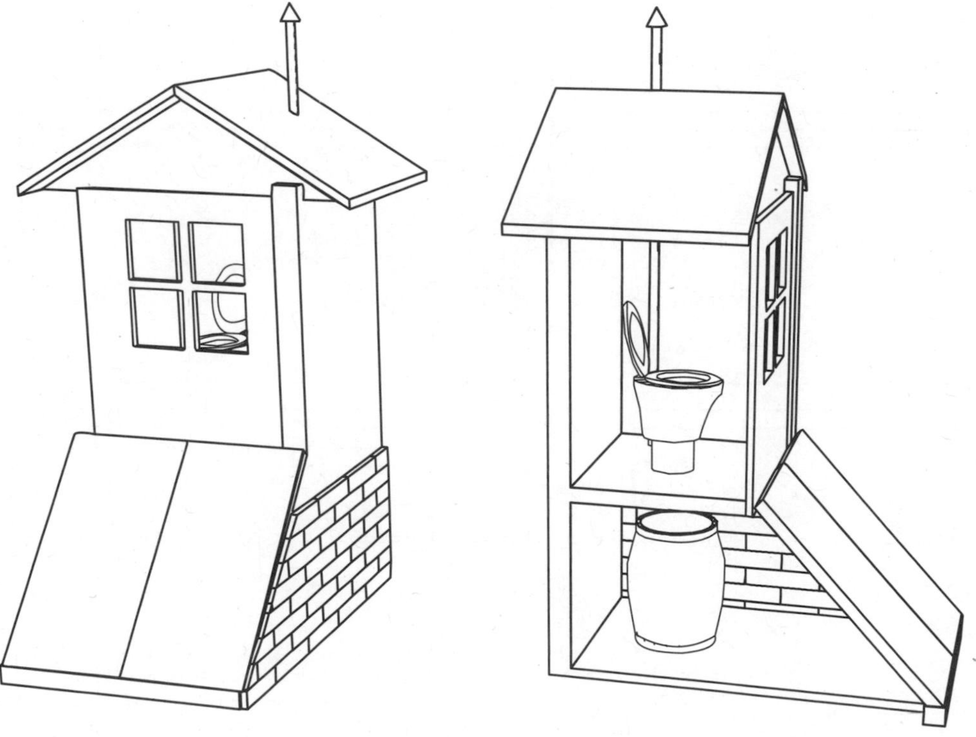

Fig. 2-11: Outdoor Composting Toilet. *This solar-heated outdoor composting toilet will perform well in most climates. "Waste" is collected in barrels in the solar-heated "basement" of the unit.*

1939, Clivus Multrum composting toilet systems have been used in homes, parks, and commercial buildings. The toilet is reliable and convenient.

Like other composting toilets, the CM relies on aerobic decomposition within a polyethylene (plastic) composting unit. The slide inside the chamber (*Clivus* is Latin for slide) perform two functions, it separates urine from feces and it delivers composted humanure to the point of removal. (Feces move down the slide, decomposing as they go. When they reach the bottom, they should be fully composted.)

According to the manufacturer, urine flows to the lowest point of the composter. As it does, bacteria convert "the chemically unstable components of urine (urea and ammonia) into a liquid end-product containing nitrite and nitrate" — a valuable fertilizer.

Fig. 2-12: Clivus Multrum. *This continuous composting toilet system has been successfully used for many years.*

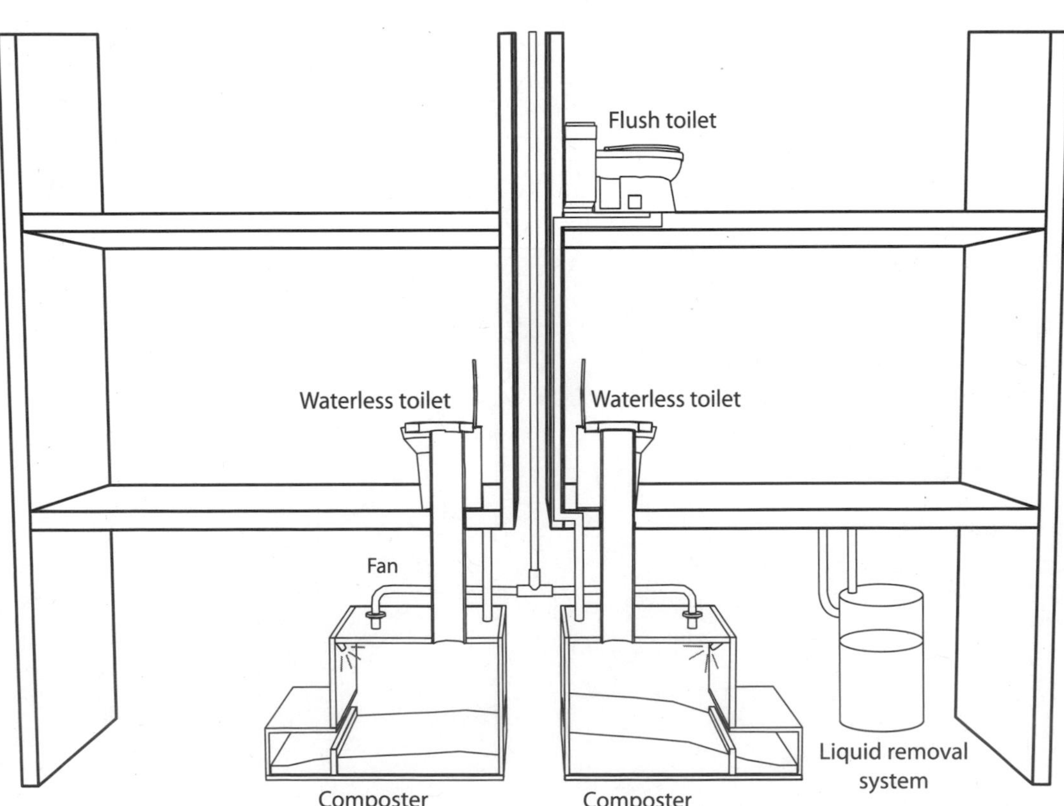

Again, according to the company, the liquid fertilizer is generated at a rate of about one gallon (3.79 liters) for every 20 uses. In most systems, it is automatically pumped out of the compost chamber into a storage tank.

The separation of urine from feces ensures that decomposing solid material — containing feces and organic additive — remain aerobic, according to the company. Aerobic conditions are vital for the function of bacteria, fungi, insects, and worms. As in other composting toilet systems, they slowly break down one's feces into humus. This process reduces the volume by more than 90%, according to the manufacturer.

In the Clivus Multrum, odors and water vapor are removed by a continuously operating fan that draws air down through the toilet and out through a vent pipe. The problem, of course, is that when installed in a home or office it also draws a lot of cold air indoors into one's home during the winter, as explained earlier.

Compost is removed after at least one year of use — and often several years. Long residence times dramatically reduce the chance that harmful microorganisms, if any, will survive. "All Clivus compost toilet systems are certified under the National Sanitation Foundation's Standard 41, as required by many state and local governments," notes the manufacturer. *Be sure any system you install meets this standard.*

Understanding How Composting Toilets Work

As any veteran composter will tell you, a compost pile needs several "ingredients" to operate properly. First and foremost, it requires organic matter — both nitrogen-rich and carbon-rich matter in the proper ratio. In a composting toilet, nitrogen comes from urine and feces. Carbon comes from feces and organic matter you add to the system such as sawdust, chopped straw, ground-up coconut shells, leaves, pine pellets, or even well-aged compost or topsoil/compost mixes. Fruit and vegetable peels or plant material like the stalks of chard add carbon-rich organic matter.

Composting toilet manufacturers also sell their own magic mix of organic additive, often made from peat moss. I avoid peat moss because it is harvested from peat bogs, naturally occurring wetlands that are home to a variety of plant and animal species. Rather than messing up natural systems, why not use other abundant organic materials such as sawdust, which is a waste product at sawmills, or commercially available compost made from cow or pig manure often mixed with urban garden and lawn "wastes."

To work properly, a compost pile also requires moisture — but not too much. The same's true of a composting toilet. You need moisture, but not too much. If moisture levels get too high, the system will turn anaerobic, and you'll have a smelly mess on your hands. As noted earlier, many commercially available composting toilets incorporate fans and heaters to maintain optimum moisture content. They also provide drains to remove excess liquid that may accumulate when the system is under heavy use.

Composting toilets, like compost piles, are sites of aerobic decomposition. So oxygen is also essential to their success. How does one ensure that oxygen levels remain sufficient in the mass of rotting humanure and organic additive?

Fans are one option. A good cover material — or bulking agent — is another, so long as it promotes *loft.* Loft is a term used in the insulation industry to describe the fluffiness of an insulation product like cellulose. Loft results from the presence of numerous air spaces in a material.

Air spaces in cover material added to humanure allow oxygen to penetrate the composting mass where it is consumed by aerobic microorganisms that digest the decomposing mass of poo.

Aeration can also be enhanced by agitation. As noted earlier, most composting toilet systems incorporate some type of mechanism to mix and aerate humanure. Sun-Mar, for instance, employs a rotating drum that permits the user to mix the contents of the composting chamber. Envirolet designed an aerator that consists of "a series of blade-like, serrated cutters that ... break up and mix the compost in the system," according to the manufacturer. Many of BioLet's systems employ an electric mixer that performs the same function.

Decomposition and destruction of potentially harmful microorganisms in a composting system also requires heat. As a youngster living in rural Connecticut, I was amazed to find steam issuing from a pile of horse manure one cold winter morning. It was then that I learned a basic fact of biology: decomposing organic material releases heat. Later in life, as I studied cellular biology, I learned that the heat I had observed came from the microbes digesting our horse's delectable pile of poo. Heat is released by microorganisms as they break down organic matter to gather up the energy they need to grow and reproduce. The cells of our bodies do the same thing. They break down food molecules like the sugar glucose to generate cellular energy (ATP). The process, however, is only about 30% efficient. Energy in glucose that's not converted into cellular energy is released as heat. This "waste heat" creates the body heat that keeps us alive.

In a compost pile, the less-than-100%-efficient extraction of energy by microbial decay also results in massive amounts of waste heat. The release of heat during the breakdown of organic matter in a compost pile — and the composting chamber of a composting toilet — ensures conditions essential to the survival of the beneficial microbes that break down organic matter. Looked at another way, you can say that the microbes create the conditions they need to survive so they can decompose organic matter. What is more, the heat they produce kills off undesirable microbes — disease-causing organisms known as pathogens that may be lurking in some people's feces.

In outdoor humanure compost piles, temperatures can easily climb to 120°F (about 49°C). At 115°F (46°C) many potentially harmful microbes, if present, begin to die. These temperatures stimulate the proliferation of a group of higher-temperature bacteria. Their waste heat, in turn, can cause the internal temperatures of a humanure compost pile to skyrocket to around 158°F (70°C). At these temperatures, pathogens have little, if any, chance of survival.

You won't see steam emanating from the composting mass inside your composting toilet because it's indoors. But internal temperatures will surely rise, and in the process kill potentially harmful microbes, including internal parasites, if any.

Further destruction of pathogens — should they exist — occurs with time. That's because good bacteria in the system produce natural antibiotics that wipe out their potentially harmful cohorts. Good bacteria even dine on nasty pathogens — that is, eat them whole. The lesson here is that the longer a pathogen's "residence time" outside the human body and in the composting chamber, the less chance it has to survive.

> **Destroying Pathogens**
>
> Potentially harmful bacteria are naturally destroyed by heat produced by aerobic bacteria in compost bins and composting toilets. They are also wiped out by antibiotic compounds released from some of the good bacteria in the system. And, they may be consumed directly by some naturally occurring microbes.

Composting toilets that contain thermostatically controlled heaters help keep temperatures high enough to maintain microbial decomposition of organic materials and potential pathogens. In a system that's operating well, the original volume of waste can be reduced by 70% to 90%, and most, if not all, of the pathogens will be destroyed before the humus is removed. Similar results can be achieved with nonelectric, or unheated, composting toilets when housed in indoors where room temperatures are relatively warm, for example, in the 70s. To be certain, I recommend processing all this "waste" from nonelectric (nonheater systems) in an outdoor compost pile for a year after it's removed from the compost chamber.

Setting Up and Maintaining a Composting Toilet

One of the most appealing aspects of commercially available composting toilets is the promise of low maintenance. It is important to note, however, that they are not maintenance-free. In fact, they require a lot more attention than a conventional flush toilet.

To ensure that they work properly, commercially available composting toilets require at least one person in the household willing to manage the system. Both wet and dry composting toilets will very likely require daily action — although daily chores are pretty simple and require very little time. Moreover, all the commercial systems require proper start-up procedures to ensure success. Follow instructions to the letter.

Start-up first requires the addition of an organic cover material or bulking agent, mentioned earlier in the chapter. It's typically made of peat moss, sawdust, wood chips, or some other fluffy organic matter. When you buy a composting toilet, the manufacturer will usually send you a bag of their own magic formula to help you get started. Manufacturers also typically supply a bacterial inoculant — a powdered mix of bacterial spores — to help "kick start" composting. These mixtures are dissolved in warm water and then poured down the toilet after adding the required amount of bulking agent.

Once the system has been set up, or primed, one member of your household (probably you, because you are reading this book!) will need to perform a cursory inspection every day or so. Daily inspections are performed in part to be sure there are no leaks. The last thing you want is for the system to leak for a week before you discover it.

Organic cover material may be added after each deposit of feces or at the end of each day. Check the manufacturer's instructions. Instructions for my Envirolet dry remote composting toilet, for instance, recommended addition of a half cup of organic matter, such as sawdust, to the toilet *every day, for each user*. So if four people are using the toilet, you should add 2 cups of sawdust.

Cover material can be dropped in from above — that is, at the throne. Or, it can be added directly to the composting chamber. I prefer going downstairs to the basement to sprinkle the sawdust directly on the feces and toilet paper deposited every day. That allows me to sprinkle the sawdust fairly evenly over the growing mound of shit. (It's not as gross as you would think.)

At the end of each week, Envirolet recommended the addition of 4 to 6 cups of cover material to ensure good coverage and adequate aeration. I aerated at least once a week, by sliding the aeration bar in and out a half dozen times. Aeration also assists in spreading the waste evenly — so it doesn't mound up. Manufacturers also typically suggest regular addition of compost accelerator — bacterial inoculant — to facilitate decomposition. I've never used an inoculant. I've always figured there are enough bacteria in the composting humanure to keep the ball rolling.

Daily maintenance takes only a minute or so, and all you will have to do is program your brain to add it to your list of things to do before hitting the sack. What is more, the world won't end if you miss a day or two.

Some manufacturers recommend against adding toilet paper to their composting toilet. I've found that toilet paper decomposes fairly readily and has never created a problem. That said, big wads of toilet paper can take a while to break down. Be sure to use it sparingly. If there's someone in your family that uses copious amounts of toilet paper, you may want to ask him or her to place the soiled toilet paper in a separate receptacle after wiping. A closed container next to the toilet works well. We dump used toilet paper into our compost pile along with kitchen scraps, weeds, and humanure from various composting toilets.

Whatever you do, be sure to read the manufacturer's instructions in the user manual very carefully, and follow them to the letter. Visit their websites for additional tips. Check out YouTube videos — anything you can find to be sure you are setting up and using your composting toilet correctly. When you purchase a system, ask questions — lots of questions — of the salesperson. Seemingly insignificant mistakes can come back to bite you!

Flies and Odors

Many people have had the distinct displeasure of using a latrine or a Port-a-Potty. The first question that comes to their minds when pondering a composting toilet is: "Will it smell?" The second question that often pops into their minds is: "Will it attract bugs?"

The answer to both questions is: "It depends." If you set up your composting toilet system correctly and use it correctly, it should be free of pesky insects and bad odors. To avoid problems, pay close attention to the instructions for installation, start up, and maintenance. Also, be sure to instruct family members to close the lid to the toilet, especially during the insect season — usually the summer.

To avoid flies, we've found that it's best to discard kitchen scraps — like peels from fruits and vegetables — directly into our compost pile, not the composting toilet. Why? Fruit skins such as orange peels often harbor eggs of tiny, annoying fruit flies. In a composting chamber of a composting toilet, these eggs will hatch and flies will proliferate. All in all, though, the insects that invade don't bite, they only tickle your butt.

To prevent other flies from entering the system, be sure to install screens on vent pipes. Always shut the door to the composting chamber after adding organic cover material.

As you would expect, all composting toilets need to be periodically emptied. How often you'll need to do so depends on the level of use, aeration, and the temperature in the composting chamber. Obviously, the more deposits one's family and friends make in a composting toilet, the more often it will need to be emptied. Warmer internal temperatures and good aeration lead to more complete decomposition. The more complete the composting, the less often you'll need to empty the chamber.

Installing a Composting Toilet

Installing a composting toilet in an existing home can be challenging. The main challenge with all composting toilets is that you'll need to install a vent pipe. As shown in Figures 2-7 and 2-13, the vent pipe allows odors to escape, typically at roof level.

In an existing home, you may need to hire a plumber to run the vent pipe through the ceiling, attic, and out the roof. Follow the manufacturer's instructions carefully. Be sure to minimize — some say avoid — 90-degree turns in the pipe. Doing so will ensure that noxious gases can escape more readily.

Bear in mind, too, that you shouldn't tie the composting toilet vent pipe into existing vent pipes in bathrooms from the sink and shower. Unless code officials state otherwise, you'll need to install a dedicated vent pipe for your composting toilet. In addition, you'll need to install a rain cap on the vent pipe top to prevent

Fig. 2-13: Caps on Vent Pipe. *Several options are available to prevent moisture from entering a compost toilet's composting chamber. Some of them, like the turbine, also help facilitate air flow, removing odors from the system.*

rainwater or snow from entering the composting toilet (Figure 2-13). In cold climates, you'll want to insulate a vent pipe running through an attic to prevent moisture from condensing on the inside of the walls of the pipe and dripping back into your composting toilet.

Manufacturers typically provide all the parts needed to install their systems except the vent pipe and vent pipe insulation. Be sure to discuss your plans in detail when talking to a sales rep to be sure they understand your installation. A good drawing is always a smart idea. That way, you can be sure they get everything right, especially the flexible pipe that connects the throne to the composting chamber of remote composting toilet systems.

When installing a remote composting toilet, you will need to cut a hole in the bathroom floor large enough for the drain pipe — usually about 8 inches in diameter. If your bathroom is tiled, you'll very likely need to remove some of the tile. Bear in mind, too, that the throne will need to be firmly attached to the subfloor — either bolted or screwed to it.

If your system includes a one-pint flush toilet, be sure that you seal the floor penetration as instructed by the manufacturer. You don't want any leakage. You'll also need to supply water to the one-pint flush toilet. If you're replacing an existing toilet, use the existing water line. If you are building anew, be sure to stub out a ½-inch cold water line to the low-flush toilet.

Composting toilets are vented using PVC pipe. Sun-Mar sells a diffuser that attaches to the vent pipe and aids in updraft — the movement of air out of the compost chamber to the roof. It also protects against downdraft in windy weather — that is, cold air blowing down the pipe into the house. Cold air forced down the pipe into the toilet may push odors into your home and could slow the rate of decomposition, at least temporarily.

For regular household use, both dry and wet systems require a drain. For our dry system, I ran a clear plastic ½-inch tube from the base of the unit (drain) into a 5-gallon bucket that contains a one-gallon clear plastic jug. It receives liquid waste that flows out of the bottom of the receptacle. (It's a mixture of urine and some dissolved fecal matter.) I often dilute the effluent with tap water and apply it very carefully to fruit trees, to the compost pile, to our gardens or to our pastures. If this is too much bother, you can run a permanent drain line to a floor drain in the basement — although you'll lose a lot of the valuable water and nutrients you could otherwise recycle.

Be sure you provide drainage, even if you operate the fan and heater a lot in a remote dry composting toilet. Otherwise, your system will leak, as mine did on two occasions in the first year of operation *without a drain.* What leaks out is a black, smelly liquid that consists of urine and dissolved feces. Very unpleasant stuff! It makes a hell of a mess on the basement floor.

Returning Nutrients to the Soil

Most states in the US require humus from composting toilets to be buried or hauled away by a licensed professional waste hauler. They also prohibit application of compost to vegetable gardens. Some individuals who promote composting toilets note that if you are going to bury it in a garden, be sure it is not one in which you grow edible vegetables. Composted humanure, they say, should only be used to supplement soils in which flowers, shrubs, and ornamental trees are grown.

Before you purchase a composting toilet, or make your own, be sure to check with local authorities to determine regulations. Don't bother the building department. Call your county health department. A veteran composting toilet user can also give you advice.

If you're not from the US, check with local authorities. Rules vary by country. Some nations are quite receptive to conditioning garden soil with humus from a composting toilet — as it should be in the States.

I can't formally recommend adding well-composted humanure to vegetable gardens, but have been doing it for years. Because it is highly likely that county officials will not monitor your actions, you can pretty much do what you want to do. If you decide to add composted manure to your vegetable gardens, be smart. Be sure humus from a composting toilet is well aged. We deposit humanure from composting toilets in a well-managed, well-contained compost pile and leave it for a full year before we add it to our garden soil. (I'll discuss compost piles for humanure in Chapter 5.) Remember, when it comes to humanure, time is your greatest ally. The longer human "waste" is composted, the fewer chances there are of any adverse health effects. The longer composted humanure resides in your composting toilet and your compost pile, the less likely you will have any problems. Be sure that it at least spends an entire spring, summer, and fall in the pile. Even then, it's best to work the well-composted humanure into the soil — and not apply it to the surface.

On a Personal Note

I have successfully buried well-composted humanure in our vegetable gardens for many years without any adverse effects. I bury it pretty deeply — about a foot down — during the off season, that is, in the fall, winter, and early spring. I have found that well composted humanure is quickly incorporated into the soil. Dig it up a month later and you'd never know I'd buried the stuff in the garden. And boy, do the tomatoes love this stuff!

Pros and Cons of Composting Toilets

Composting toilets are designed to create a user-friendly way of recycling the nutrients in human excretions. Most brands are attractive, and they generally work well in cabins, cottages, workshops, and homes — if maintained properly.

Commercially manufactured composting toilets are designed to reduce human exposure to fecal microorganisms in freshly deposited feces and kill potentially harmful microbes, including parasites. Composting toilets on the market today are also designed to either passively or actively evaporate water from feces and urine, and they come with a drain port to which you can attach a drain line to remove excess liquids. I always recommend installing a drain, just to be sure. I've found that the commercial composting toilets aren't as watertight as I would expect them to be.

On the downside, self-contained composting toilets are quite large and take some getting used to. They'll definitely change the aesthetics of your bathroom. Remote composting toilets are much nicer, though you'll need to have a relatively warm basement to house the composting chamber.

Composting toilets require care and attention. As noted, that detailed attention usually falls on the shoulders of one family member. He or she periodically must add organic bulking agent/cover material and mix the contents of the receptacle to facilitate aeration (oxygen penetration) and bacterial decomposition. The caretaker must also monitor liquid drainage, and periodically empty humus from the unit into the compost pile. The caretaker must also maintain the compost pile.

To avoid problems, be sure to purchase a unit that's appropriate for your use — or better yet, one that's a bit larger than you think you'll need. If your home is flooded with guests several times a year, you may want to have a backup toilet, perhaps a sawdust toilet described in the next chapter, for use during such times. Gentleman can be asked to urinate outdoors unless you live in a city or suburbs or your guests would be offended by such a barbaric act.

On a Personal Note

Over the years, I have found that wet and dry composting toilet systems by Sun-Mar and Envirolet (respectively) can be fairly easily "overwhelmed" by flush water and urine, respectively. When they do, they'll leak. The low-flush Sun-Mar I use in the classroom is particularly vulnerable to leakage. If there's any stoppage in the drain, liquid seeps out onto the concrete floor of my wife's business, Evergreen Naturalworks, located below the classroom building. She's a good sport about it, but I know she's not happy when this occurs.

If you are installing a dry composting toilet in your home for daily use or in a cottage or cabin where it will be used for more than a week at a time by a handful of people, be sure to install a drain to remove excess liquids. Install a bomb-proof drain line and monitor the unit carefully.

Conclusion

Commercially available composting toilets provide a relatively simple, though rather expensive, way to enter the world of responsible nutrient recycling. If you are constrained by budget or are a do-it-yourselfer, don't despair, there are lots of other ways to achieve the same goals. I'll describe one of them, the sawdust toilet, in the next chapter.

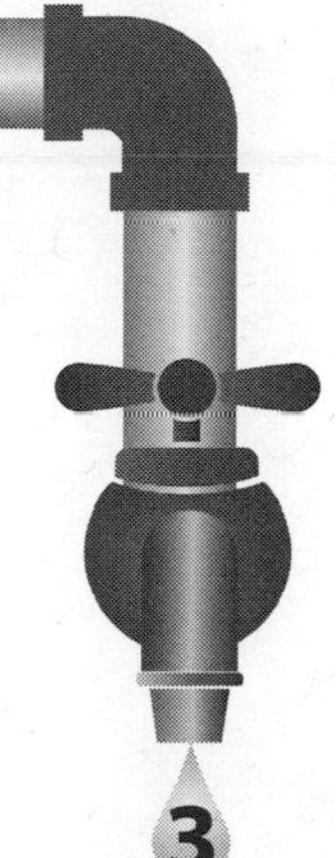

Sawdust Toilets: Simple, Effective, and Intelligent

One of my favorite composting toilet systems is the simple sawdust toilet popularized by Joe Jenkins, author of *The Humanure Handbook*. Pooping in a bucket of sawdust, then carting the contents to a backyard compost pile every few weeks may seem primitive and out-of-step with modern toiletry. That said, depositing excrement into a water-filled porcelain receptacle isn't that much more glamorous, either, if you think about it. Let's face it, defecation is not one of humankind's most exalted activities.

The advantages of the sawdust depository over a standard flush toilet, however, are many. The system uses no water, except to rinse buckets from time to time. And, it allows us to capture the nutrients in urine and feces and, even more important, to put them to good use. In contrast, the modern flush toilet is extremely wasteful of precious resources: clean drinking water and inorganic and organic soil nutrients — vital to soil productivity and continued food production.

The Sawdust Toilet

Like many things in life, the simplest ideas — even some of the most primitive ideas — like sawdust toilets — are sometimes head and shoulders above the best modern technology has to offer (Figure 3-1). In fact, I have used many types of toilets over the years — smelly outhouses at National Forest campgrounds, nauseating Port-a-Potties at county fairs, even grosser "honey buckets" on long kayak trips through the Grand Canyon, fancy composting

Fig. 3-1: Sawdust Toilet. *The sawdust toilet is simple to build and easy to use. It will allow you to capture all the nutrients in your feces and urine. These toilets are odor-free if maintained properly but require an outdoor compost pile. We've used a sawdust toilet in our Colorado and Missouri homes and in our classroom at The Evergreen Institute for many years.*

toilets, modern flush toilets, and simple sawdust toilets. My favorite of all — the one that is easiest to use and helps me accomplish my goals of recycling nutrient-rich "wastes" — is the sawdust toilet.

Truly, sawdust toilets are one of the easiest, cleanest, and most convenient options I've used for capturing and recycling human solid and liquid waste. In fact, I've had much better luck with my simple, inexpensive sawdust toilet than my more sophisticated and considerably more expensive commercially available composting toilets. I continue to use my sawdust toilets to this day — either when I'm having trouble with one of my commercial composting toilets or when we have a house full of guests and I want to be sure that we don't overwhelm our remote composting toilet with urine. Truth is, we're using a sawdust toilet right now in my educational center and office to demonstrate one of the composting toilet options and, of course, to help reclaim the nutrients in excrement and urine left by me and the students who enroll in my workshops.

Even if you think that a sawdust toilet might not pass muster in your home, you may want to make one anyway for emergencies. They're not only ideal for disasters like week-long electrical outages that follow

Table 1
Possible Uses for a Sawdust Toilet

Possible Uses
New home construction (worksite toilet)
Marine or boat toileting
Campsite and/or rv toileting
Disaster planning and civil defense toileting
Low-cost toilet additions to apartments, townhouse, mobile home, condo, garage, workshop, office, barn, studio, or outdoor location
Temporary toilet for use during bathroom remodel or construction
Family emergency/disaster backup plan
Fallout shelter or storm shelter toileting
Toileting solution for location with no water or sewer installation

Source: The Sawdust Toilet Company (www.sawdusttoiletcompany.com [defunct])

in the wake of hurricanes, they're perfect for remote cabins and cottages and even for use on boats. They're great for workshops and for worksites, and make a cheap and convenient second or third toilet for a home. If you have a storm shelter, it wouldn't be a bad idea to have a sawdust toilet and necessary supplies standing by — just in case. Table 1 lists these and a host of other possible uses for sawdust toilets.

Handy and relatively easy to use, sawdust toilets can be made to be quite attractive. Yes, attractive. The Internet is peppered with photos of innovative and, dare I say, rather pleasing designs.

Just what is a sawdust toilet? How do you make one? How do you use it and how do you compost and safely recycle the waste? Should you be concerned about disease? These are some of the questions I'll answer in this chapter. I'll also discuss the pros and cons and some of the potential problems you might face when using one.

If you're living your life consistent with the goals of sustainability and self-sufficiency, you understand the logic of sawdust toilet use — or any composting toilet for that matter. Nevertheless, convincing skeptical family members to use a sawdust toilet may be a challenge. I don't think my boys ever went near mine!

Anatomy of a Sawdust Toilet

Sawdust toilets are part of a strategy known as a *cartage system.* Cartage system simply means that you cart your urine and feces out of the house in a bucket — or some other container. It's not as horrible as it sounds, though. Urine and feces are covered by sawdust or a similar material after each use. Sawdust absorbs liquids almost completely and — quite surprisingly — kills virtually all odors. Feces quickly dissolve in the sawdust-filled bucket, so when you deposit them in a compost pile, you won't be seeing any remnants of your latest greatest bowel movements.

Bear in mind, a sawdust toilet is not a composting toilet. When combined with an outdoor compost pile, it becomes part of a system that, if used intelligently, permits us to safely compost excrement and capture the many inorganic and organic nutrients contained in our daily excretory "wastes."

Some online sources recommend disposing of the contents of sawdust toilets in tightly sealed plastic bags along with your trash. Although expedient, this approach negates the purpose of the sawdust toilet. What is more, it is illegal.

Yes, technically it's illegal to dispose of human excrement — even soiled diapers — in landfills. That's because feces and urine could potentially leak out of landfills into groundwater. I realize that millions of parents dump their children's dirty diapers in the trash every day, but that doesn't make it right — or legal. Authorities just look the other way.

For the ecologically minded, depositing urine and feces in landfills, septic tanks, or sewage treatment plants is a waste of highly valuable resources — resources that should return to the soil.

For the ecologically minded, depositing urine and feces in landfills, septic tanks, or sewage treatment plants is a waste of highly valuable resources — resources that should be returned to the soil.

Figure 3-1 shows a simple sawdust toilet. I use the adjective "simple" because this option consists of nothing more than a wooden box with a hinged lid to which a toilet seat is attached. Inside the box is the guts of the unit, a five-gallon plastic bucket. For ease of use, you'll probably need four of them. I'll explain why shortly.

Sawdust toilets are easy to build, even if you're only moderately handy. You'll need a some plywood, some short pieces of 2 x 2, one-inch wood screws, a skill saw, a jig saw, a cordless drill, an appropriate screw bit, and a toilet seat — and that's about it. Because there are plans and YouTube videos online that show you how to build a sawdust toilet, I did not include detailed instructions here. Figure 3-2, however, provides the details you'll need to purchase materials, cut wood, and assemble a sawdust toilet, including key measurements for those who are interested in building their own.

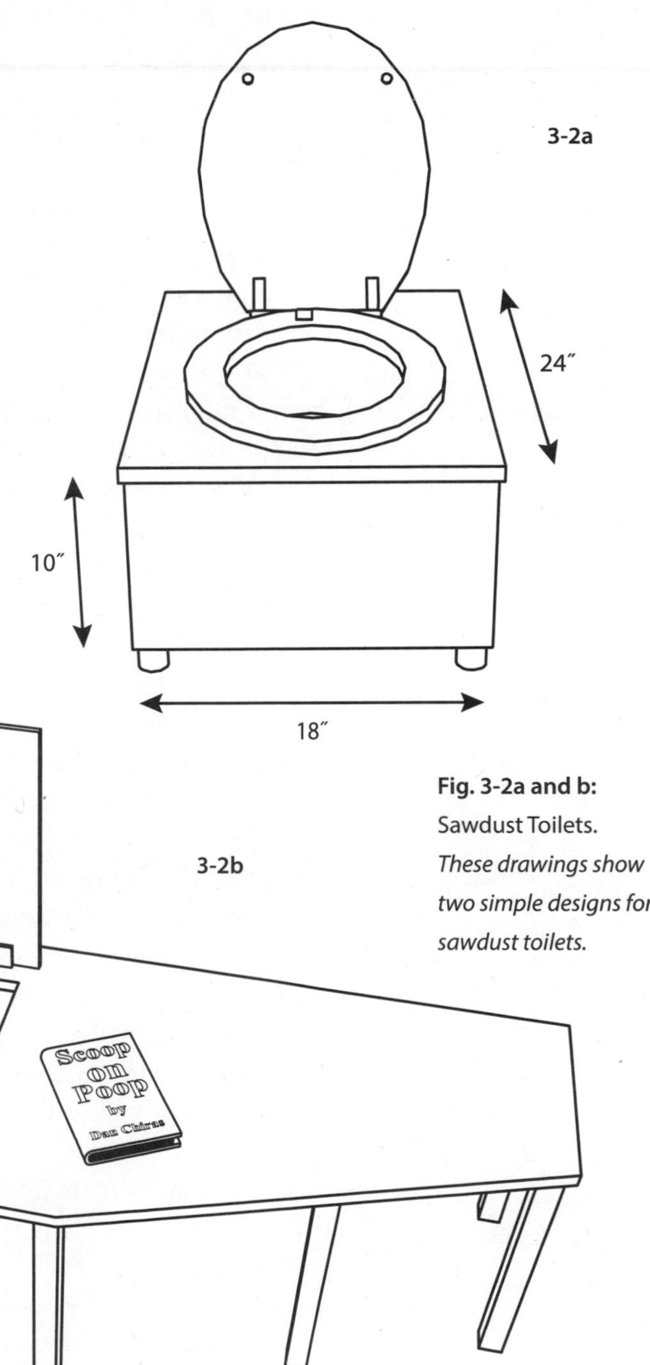

Fig. 3-2a and b: Sawdust Toilets. *These drawings show two simple designs for sawdust toilets.*

How Do Sawdust Toilets Work?

Sawdust toilets are extremely simple to operate and maintain. When Mother Nature calls, you open the toilet seat, and then do your thing. You can either sit or stand, depending on your sex, immediate needs, and preferences. You'll figure all that out in quick order. You deposit urine and feces in the bucket, then deposit the toilet paper, if any, in the bucket, and then cover the "waste" with a handful of sawdust, fine wood shavings, or some similar material sprinkled on top. Once that's done, and your pants are buttoned up, you close the lid and go about your other business, knowing you're helping save the planet one bowel movement at a time. (I couldn't resist that!)

The success of a sawdust toilet depends on proper cover material — and proper application thereof. Cover material keeps flies out and contains odors. It's amazingly effective at both. If you keep the "waste" covered at all times, your sawdust toilet should be virtually odor-free. In many years of using a sawdust toilet, I've found that they truly live up to their odor-free reputation. The only problem I've ever had was when I used wood shavings as a cover material or failed to fully cover feces.

Cover material is stored in a nearby container. A clean, five-gallon plastic bucket with a lid works well, though you can create a more attractive container if you'd like to spruce up the loo a bit. You'll also need a large scoop — or perhaps a large plastic yogurt container — to transfer the cover material from the bucket to the sawdust toilet.

For best results, I recommend using sawdust, ground coconut fiber, or fine pine wood shavings as cover materials (Figure 3-3). They'll also serve as a bulking agent that provides loft to promote aerobic decomposition. I've experimented with many different cover materials over the years, including pine wood shavings and pine pellets. The latter are sold as bedding for horses or as fuel for pellet stoves. I have found that wood shavings don't cover fecal material very well, unless you add a ton — and then if you do, your buckets will fill up very quickly. Pine pellets just out of the bag do an even poorer job of covering poo and containing odors.

Pine pellets absorb urine well, but cover feces very poorly. You have to add a lot to fully cover a bowel movement. Pine pellets also swell up and fill the bucket pretty quickly. That said, pine pellets are inexpensive and widely available. What is more, they can be easily converted to sawdust. All you need to do is add a little water.

Fig. 3-3: Cover Materials. *The top row shows dried leaves, straw, and coarse pine shavings. They work well in remote toilet compost chambers especially homemade ones and outdoor compost piles. The middle row shows pine pellets, coconut coir (ground up coconut shells) and fine pine shavings. Pine pellets are not recommended unless they are first converted to sawdust. Coconut and fine pine shavings work well in sawdust toilets and composting chambers of composting toilets. The front row shows sawdust from pine pellets and well-aged compost and cow manure. These options work well in sawdust toilets and compost chambers.*

Pine pellets can be purchased at farm supply stores, feed stores, and some home improvement centers (as fuel for pellet stoves). I've used them a lot because they're much easier to acquire than sawdust. By the way, my wife and I have also found that pine pellets make an extremely effective, inexpensive, and compostable kitty litter. In fact, they're the best kitty litter we've ever used! When they become soiled we apply them to our fruit trees as mulch.

To convert pellets into sawdust, pour them into a plastic 5-gallon bucket. Fill about one-third of the way. Add water a little at a time. Let it stand for a few minutes, then stir the mix with your hands or an appropriate implement. If pellets still remain after five or ten minutes, add a little more water, and stir again. Once all the pellets have absorbed the water and turned into sawdust, add more pellets. Add a few more cups of water and stir. Let the mass sit for a while. You'll find that a half a bucket of pellets, when wetted, will expand to about three-fourths of the bucket.

I store dampened pine pellet sawdust in a bucket next to our sawdust toilets and sprinkle a handful after each defecation. You can also use a cup or a scoop to transfer the sawdust.

While I use pine pellets (for reasons noted above), I've found that most any organic matter will suffice as cover material in a sawdust toilet. Sawdust has performed best, but even dried, crumbled leaves; fresh or dried grass clippings; potting soil; or topsoil mixes containing well-composted animal manure and organic compost can be used.

I've successfully used commercial compost mixtures consisting of well-aged cow manure and compost (Figure 3-4). This product and other similar products are sometimes sold as topsoil or as an alternative to conventional potting soil at home

Convenient Cover Material

Keep your container of cover material close to the toilet and provide a scoop to transfer the cover material. I kept my cover material bucket in a bathtub next to the sawdust toilet. At first, I transferred cover material by hand, but found this to be quite messy. To reduce spillage, I converted to a large scoop. A 24-ounce sour cream container worked well, too. Only fill whatever you use part way to avoid spilling.

Fig. 3-4: Topsoil in a Bag. *Dried a little, products like this work extremely well. They can be mixed with sawdust, dried leaves, and chopped straw, to name a few.*

improvement centers and nurseries. These mixes look, feel, and smell like a very rich top soil. Right out of the bag, however, they contain a fair amount of moisture. To ensure that it spreads evenly over newly deposited "waste" in our sawdust toilet, I dry it out a bit first. I place mine in a 18-gallon plastic tub (tote) in the garage in the summer where it dries out fairly quickly with periodic mixing. (Mix the deeper layers with the superficial layers to accelerate drying.) I have also mixed this material with sawdust and other organic materials like dried leaves. These mixtures works great, as they probably add microorganisms that speed up the decomposition of the sawdust and feces.

The master of the sawdust toilet, author Joe Jenkins, uses sawdust from sawmills. He calls it "rotted sawdust" in his popular book, *The Humanure Handbook*. Although the term "rotted" may turn people off, what Joe means is that the sawdust he uses comes directly from a sawmill and is a bit moist. He purchases sawdust by the pickup load, then stores it outside, where it is exposed to the rain and snow. He periodically transfers some indoors in a plastic bucket for use in his sawdust toilet. According to Jenkins, wet and partially decomposed sawdust helps accelerate the composting process.

Sawmill sawdust is a great option and extremely inexpensive. In fact, sawmill owners may be willing to let you have it for free — if you haul it away. You can shovel it into the back of a pickup truck, but you don't need a pickup truck to secure sufficient quantities. You could carry it home in the trunk of sedan or the back of hatchback in garbage cans, 18-to-30-gallon totes, or sturdy plastic bags.

If you live in a forested region, like the eastern half of the United States and Canada, the Pacific Northwest, and many parts of Europe, you'll very likely find that there is a small sawmill nearby. There's one within ten miles of my house in east-central Missouri. When I lived in the foothills of the Rockies in Colorado, there was a sawmill within the same distance.

If you live in the Great Plains states or in a city or suburb, don't despair. You may be able to obtain sawdust from a local wood-worker. Some sources advise against the use of sawdust from cabinet makers and other wood workers. They say it is too dry.

While sawdust from wood-working shops comes from kiln-dried lumber and is much drier than sawdust from sawmills, once deposited in a toilet, it quickly absorbs moisture and will remain wet in a sawdust toilet and a well-maintained humanure compost pile. Also, bear in mind, that very little — if any — composting occurs in the sawdust toilet. Most of the decomposition occurs in the compost pile.

Urine and feces in a sawdust toilet create a bacteria-rich organic mass. When added to a compost pile in the spring, summer, and early fall, it decomposes pretty quickly. Decomposition may be accelerated if you sprinkle in a little topsoil or soiled animal

Why Avoid Sawdust from Pressure-treated Lumber?

Pressure-treated lumber (PTL) is used to build decks, docks, and fences — applications where wood is in fairly continuous contact with moisture. PTL resists microbial decomposition in a wet environment. Its resistance to microbial action is the result of toxic chemicals that are driven into the wood under pressure. The chemical compounds in pressure-treated lumber are not only toxic to bacteria and other microorganisms, they are also toxic to insects such as termites. As a result, they help prevent decomposer organisms large and small from "digesting" wood, causing it to decay. Several types of pressure-treated lumber are available.

In earlier years, pressure-treated wood was saturated with a water-based solution containing chromium, copper, and arsenic (CCA) — chemicals that are toxic to microorganisms. It was widely used for decks, docks, landscaping lumber for raised bed gardens, and sill plates over concrete floors. CCA lumber is now restricted for use only in commercial projects, primarily for poles, pilings, and bridge timbers.

The most common product in use around our homes today contains a preservative that consists of copper formulated with an ammonia or amine compound. It's still toxic to microbes, but supposedly safer to use in residential applications.

PTL is amazingly effective. I've pulled pressure-treated posts out of the ground that have been in place for many years on my farm, and they look like they're brand new.

It should go without saying: because you want to promote decomposition of humanure and cover material/bulking agent by microbes and insects, it's best to keep toxic chemicals that thwart biological decay out of the mix. It's pretty much common sense, eh?

bedding (like straw) over the compost pile. Remember, feces- and urine-drenched sawdust will reside in the outdoor compost bin for a full year, maybe two years, so there's really no need to worry about whether the sawdust you use in your toilet is dry.

Whatever you do, be sure *not* to use sawdust from cedar, redwood, or other naturally rot-resistant woods or from pressure-treated lumber. Although cedar shavings or sawdust produce a very pleasant smell and will help mask odors, they (and other naturally rot-resistant woods) contain oils that resist microbial decay. These oils are natural preservatives. That's why cedar fence posts resist rotting even when buried in the ground for decades. In a sawdust toilet system, they will very likely slow down the rate of decomposition.

Pressure-treated lumber contains toxic chemicals that resist breakdown as well (see accompanying text box). Fortunately, most wood-workers don't use pressure-treated lumber. It's primarily used for building decks and docks and for sill plates over concrete floors. It's cut onsite, outdoors, and not typically collected.

How Often Will You Need to Empty the Bucket?

When I first started using a composting toilet, I was amazed at how quickly the bucket filled up — even with only one of us in a household of three using it. Being the sole depositor and working fulltime at home, it took between 10 and 14 days, sometimes a bit longer, to fill. With two adults using the sawdust toilet, it fills nearly every week. A family of four may need to empty the bucket every two to three days.

How often you'll need to empty your buckets varies not just with the number of people using it, but with their daily food consumption, how much toilet paper they use, and how much cover material they sprinkle over their deposits. Heavy eaters produce a lot more fecal "waste" than lighter eaters, which will necessitate more frequent emptying. Extra generous depositions of cover material/bulking agent and toilet paper will also reduce the time between trips to the compost pile.

A lot of sawdust toilet users keep two spare, clean buckets on hand — in the garage or in the basement, for example. When the bucket in use fills, they place a new bucket into service. They put a lid on the recently filled bucket, and then set it aside in a safe place. After the second bucket fills, they cart the two filled buckets to the compost pile. The third bucket is then placed into service. This system reduces maintenance and is easier on the body. It's much easier to carry two buckets at a time rather than one (Figure 3-5). It is considerably more balanced and creates less strain on the body.

Fig. 3-5a and b: One Bucket vs. Two. *As you can see in these photos, carrying two buckets results in a more balanced stance.* Credit: Linda Stuart

Pros and Cons of Sawdust Toilets

It's no secret, I am a huge fan of sawdust toilets. (That statement just killed my run for the presidency in 2024 — as if I had a chance!) Sawdust toilets allow me to capture *all* the nutrients from my family's daily excretions, and, even more important, put them to good use. They are inexpensive to build and operate. They use very little water. (The only water needed is to rinse out the buckets between uses.) They should never leak, and unless you're a complete dolt, they should never overflow. If you maintain them properly, they'll operate odorlessly. What is more, sawdust toilets help reduce our carbon footprint by reducing the energy consumption required to pump water, treat waste and transport sludge to landfills. They also reduce toxic chemical use (chlorine) at sewage treatment plants.

As already pointed out, sawdust toilets help us return nutrients to garden soils. This, in turn, reduces the need for fertilizers. Incorporated into soils and then into fruits and vegetables grown in our backyards, composting toilets help us build a more self-sufficient and sustainable means of food production. In so doing, they help us reduce the amount of energy required to transport food to our tables and the amount of pollution generated by the combustion of transportation fuels. And, lest we forget, they help us put our ethics into action. For those inclined toward a sustainable future, sawdust toilets are one of many simple technologies that will help us operate by the rules of nature.

For those inclined toward a sustainable future, sawdust toilets are one of many simple technologies that will help us operate by the rules of nature.

The downside is that they require frequent emptying, especially if you have a large family. The contents of your sawdust toilet must be hauled to the compost pile even in the dead of winter. In addition, sawdust toilets may not be popular among all family members, relatives, and guests. And, they may get you in trouble with city officials or neighbors. They may smell a little — if not properly operated. And, they require an outdoor compost pile. City dwellers will have a tough time incorporating a sawdust toilet into their lives, as might suburbanites.

For those who are squeamish but still in tune with the ecological mandate, this technology requires one to get down and dirty with his or her daily excretions. Someone in your family will need to take responsibility for regularly dumping the nutrient-rich mass and maintaining an outdoor

My Experience with Nature's Head Composting Toilets

I have used a Nature's Head composting toilet since January, 2011, in my office bathroom — and continue to use it today. I first put it into service when my house in Missouri burned to the ground. At that time, I was forced to move into my office in the classroom building at The Evergreen Institute. I set up the toilet in one of our two bathrooms to see how well it worked. Over all, I liked it a lot, although there are some downsides, which I'll get to shortly.

composting pile. If you have ever changed a dirty diaper, though, I think this process will prove much more pleasant. You should never get peed on. If you have a cat and scoop out its litter box, you'll also find the sawdust toilet to be much more agreeable. Fresh cat poo has to be one of the most disgusting smelling things on planet Earth. If you pick up dog poo in a bag when you walk your pooch, you'll find the sawdust toilet is a walk in the park!

Nature's Head Toilet

You can make your own sawdust toilet for $20-$40. Or, you might want to consider another option that I've used with great success as well: Nature's Head (Figure 3-6). This unique toilet system will set you back $925 plus shipping (at this writing). Nature's Head, when used regularly, is a lot like a sawdust toilet with a couple distinct differences that I'll talk about shortly.

Fig. 3-6: Nature's Head Toilet. *Students and I use this toilet located in my classroom building at The Evergreen Institute. To the left is a bucket of sawdust made from pine pellets we purchase from Tractor Supply.*

What is a Nature's Head Composting Toilet?

Figure 3-6 shows a Nature's Head composting toilet. It is a self-contained unit made from rugged plastic.

Unlike the sawdust toilet, Nature's Head separates urine from feces. Feces are deposited in a chamber immediately behind the urine repository. Feces drop through a hole in the bowl. When not in use, it is covered by a trap door to prevent odors from escaping. The trap door is opened manually using a lever on the side of the toilet. (Be sure to remind all users to open the trap door before dropping a bomb! Otherwise, you'll have a mess on your hands.)

When someone is seated on the toilet, urine flows naturally into the two-gallon opaque plastic container located in the front of the unit. Men can also urinate into the toilet when standing, but need to aim carefully to prevent urine from splashing on the closed trap door that leads to the chamber where feces are deposited. As a caution, ladies, be sure not to sit too far back on the unit!

With careful use, feces and urine should never mix. If urine enters the rear chamber, be prepared for odors. The combination of the two is well ... not very pleasant. Heck, let's be honest, it's disgusting!

As in a sawdust toilet, cover material is sprinkled over feces after each use. I've found that all the cover materials/bulking agents that work well in sawdust toilets perform well in my Nature's Head toilet. Users also need to periodically mix the sawdust and fecal matter. This is accomplished by rotating the handle on the side the unit. I usually agitate mine after every use while I'm using the toilet. It gives me something to do. Mixing helps blend the organic cover matter/bulking agent with nutrient-rich fecal matter — and could help accelerate composting.

Are They Really Composting Toilets?

Although Nature's Head toilets are referred to as composting toilets, I've found that with regular daily use, they really don't deserve that title. They're really just fancy sawdust toilets. That's because, in my experience, they fill too quickly — too fast to allow any appreciable amount of composting.

With daily use, Nature's Head's "composting" toilet can be used as part of a cartage system — a repository for those nutrient-rich excretions we all produce with daily regularity. If the unit is used infrequently, say for a few weeks at the beginning of the summer in a cabin or cottage, then again later in the year, some decomposition (composting) of the cover material and fecal matter will invariably occur. Even so, the folks at Nature's Head still recommend composting the output in a compost pile to give it time to fully decompose. As you will learn in the last chapter, additional time spent in a properly managed compost pile is time well spent. It ensures adequate destruction of potentially pathogenic microorganisms.

Getting Started

To prepare this toilet for use, the manufacturer recommends first depositing peat moss (without any additives such as fertilizer) or some similar organic material, such as coconut fiber or sawdust, in the poop chamber. Two one-gallon freezer bags full of pre-moistened sphagnum peat moss or coconut fiber should suffice. According to the manufacturer, the slightly moist organic material should rise to the level of the agitator crank handle. "The peat moss should be damp and crumbly," they recommend, "never wet or soupy." You can begin using the toilet at that point. Just be sure to mix the cover material and feces after every use. If you start with a half-full chamber, you will very likely not need to add cover material/bulking agent for a few weeks, depending on the amount of use.

In the five years I've used my NH toilet, I have found that an initial deposit of a much smaller quantity of cover material works quite well. I add four or five cups of cover material when starting a new cycle, then add a handful each time I use it. Adding the manufacturer-recommended amount of sawdust (up to the agitator handle) made it difficult for me to turn the agitator.

Emptying the Toilet

Nature's Head toilets are fairly easy to empty. When the time comes to empty the urine container, you simply flip the latches on the

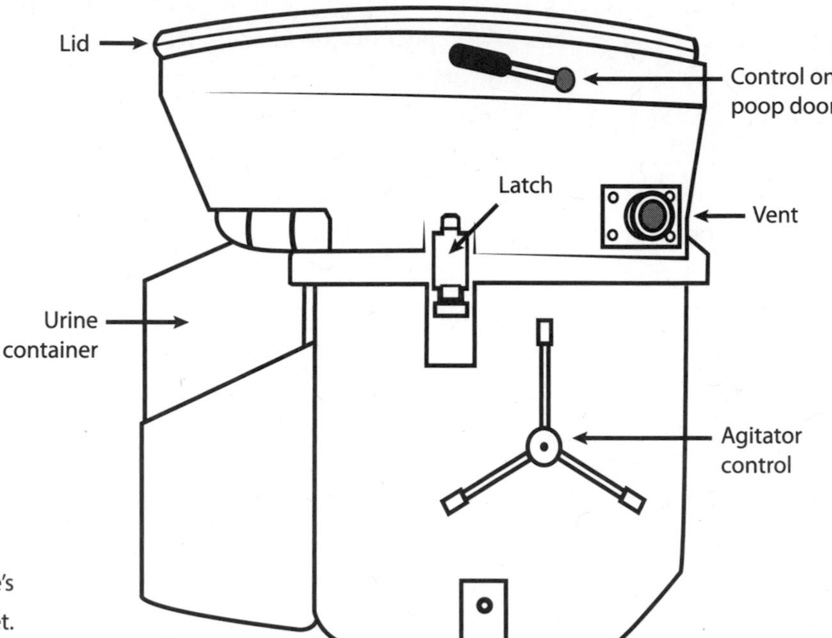

Fig. 3-7: Detail of Nature's Head Toilet.

sides of the toilet, shown in Figure 3-7. Next lift the upper section, which is hinged in back. This allows you access to the two-gallon urine container. Place the cap on the plastic container (provided when you purchase the toilet) and lift it out of its holder. Urine can be deposited directly into a compost pile or diluted and used to fertilize vegetables, flowers, ornamental trees, fruit trees, or berry bushes. It goes without saying, but I'll say it anyway, be careful not to spill it on yourself; urine can get pretty smelly.

On our farm, we apply the urine to our pastures, targeting areas that need fertilizer. I spread each container over a four-by-four-foot area to prevent overdosing the grasses with this nitrogen-rich fertilizer. It is difficult to prevent splatter, so pour carefully. It's not the most pleasant task in the world, but it is rewarding to see all those nutrients being returned to the soil to nourish healthy pasture grasses.

Emptying the poop tank is a bit more complicated. If you have hooked up the vent, which I'll discuss shortly, you will first need to detach the 12-volt power connector that powers the fan. You then slide the upper portion of the toilet (the seat) off its open-ended hinge at the back of the unit. With the urine container removed, the poop chamber is carted to a humanure compost pile, discussed shortly. Figure 3-8 shows the toilet completely disassembled.

Nature's Head toilets are designed primarily for remote cabins, boats, and RVs. Several options for disposal are possible. "The contents of the solid waste tank may be safely placed into a conventional dumpster if it ... is bagged and sealed," the manufacturer recommends. "The unit is sized so that a kitchen garbage bag will go over the end of the main tank. Simply flip it over and you're done." I prefer to compost the "waste," as I suspect most readers will.

In the spring, summer, and fall, I cart the nutrient-rich materials into our compost pile. "Because there are no liquids in the main tank, even full, it is never heavy," asserts the manufacturer. I agree, but even though it is not very heavy, it is fairly awkward to handle. I'd recommend rigging up a handle to cart the poop-filled

Fig. 3-8: Nature's Head Toilet Disassembled.

Fig. 3-9: Hauling Poo. *The author hauls the poop chamber from his Nature's Head Toilet to his compost pile. His advice is to use a wheel barrow or a wagon or devise a handle so you don't need to carry it against your body.* Credit: Linda Stuart

chamber to the compost pile. Figure 3-9 shows me carrying the unit *sans* handle. As you can see, it is not terribly pleasant. Alternatively, you could transport it in a wheelbarrow or wagon.

In the dead of winter, I often bury the tank contents in the raised beds in our greenhouse. Burying the sawdust/manure mix 6 to 12 inches down results in fairly rapid decomposition. Within a few weeks, it is fully incorporated into the soil — and boy do the tomatoes love that soil!

The manufacturer notes that "there is no need to clean the tank when you empty it; any leftover matter will help start the composting process all over again."

If you want to promote composting in the poop chamber and make dumping it out a bit more pleasant, you might want to consider purchasing an extra base. When the first base fills, remove it and place a lid on it, then set it aside in a safe place to continue to compost. If it can sit for a month or two in a place that's warm, the sawdust or peat moss and poop travel further along the road toward humus. "The extra base comes complete with all the necessary hardware, agitator, and bottle holder. It also comes with a vented lid so the contents can be set aside and allowed to compost," notes the manufacturer. Alternatively, you can dump the waste into a plastic bag that can be placed in a storage bin or bucket, perhaps a 5-gallon bucket. As long as the lid is ventilated, composting should continue. Surprisingly, the bucket releases almost no odor.

Installing a Nature's Head Toilet

Installation of Nature's Head toilet is pretty easy. The unit sits right on the floor and can be secured by two L brackets. You don't need a lot of room, either. Be sure there's enough headroom to allow family members and friends to maneuver into position, though.

If you install a system and find that it releases odors, you can retrofit a fan and vent. They're extras. I installed the toilet in our energy-efficient classroom bathroom that has 9-inch-thick walls, but I never installed the power vent. I didn't want to compromise the air-tightness of the building and, fortunately, found that the unit is remarkably odor-free. That said, the manufacturer does recommend venting to "keep your bathroom smelling fresh."

Venting may also help facilitate the growth of aerobic bacteria required for composting. While oxygen helps facilitate decomposition, as I noted earlier, don't count on much composting occurring inside one of these units if it is being used regularly.

Installing a vent can be difficult, especially if you aren't used to poking holes in walls, ceilings, or floors of buildings. If you are unsure about the proper procedure, hire a handyman or woman to help. When doing it yourself, be careful when cutting a hole in a wall; you don't want to damage electrical wires or water pipes that run inside the wall. And you don't want to damage the walls of an RV or boat, either. So get help, if you are unsure.

For installation in a home, cottage, cabin, workshop, or any other permanent structure, 1¼-inch PVC vent pipes can be run horizontally through an exterior wall. You can then install a 90 degree elbow, turned down to prevent wind and rain from blowing in. Be sure to seal the penetration very well with a high-quality silicon caulk or, better yet, a foam spray, to prevent infiltration and exfiltration. Also be sure to attach some fine-mesh window screen over the opening to prevent bugs from entering. I'd use plastic window screen. Or, if you want, you can purchase a screened vent from the manufacturer. (They offer a couple options.)

Excellent instructions are provided online for installation, so I'll leave you to figure out the rest on your own.

Pros and Cons of Nature's Head Toilets

I like this system a lot, although not as much as my sawdust toilets. It has worked well for us and doesn't smell much. I found it easy to use, although visitors are reluctant to use it. They typically prefer the one-pint flush toilets in our classroom building or the dry remote composting toilet in our home.

Installation without ventilation is a snap. In fact, I didn't even secure my Nature's Head toilet to the floor. I just placed the unit on the floor and began using it. The toilet has served me well — and continues to be my toilet when working in the office. It and a sawdust toilet have helped me capture a ton of valuable nutrient-rich excretions that have been safely returned to the soil.

One problem I found was that the urine container fills quickly when serving two or more people. Even though overflow is supposed to spill into the container base (bottle holder), which has a fairly significant reserve, we have found that urine backs up in the pipe that connects the urine diverter up top with the urine bottle down below. As a result, urine spills all over the floor when I lift the top section up to remove the urine bottle if it has overfilled. The lesson in all this is to watch urine levels in the bottle very carefully. Empty it frequently or you'll be mopping up smelly pee way more than you would prefer. (When we do need to mop up anything, we use paper towels that get composted and an earth-friendly cleaning agent.)

I have also found that minerals and salts build up on the inside of the urine bottle. I rinse them out after every cycle. To do so, fill the recently emptied bottle with tap water, then shake it vigorously. To dislodge the deposits. You can pour them down the drain or add them to your compost pile.

Another problem I had was that the seal between the upper section and the base came off within the first year. Without the seal, flies invaded. In the summer, they laid eggs in the humanure, and the unit was soon swarming with larvae and flies. To kill the flies and larvae, I sprinkled some diatomaceous earth over the poop/sawdust mixture (advice from the manufacturer). This worked remarkably well.

The final downside of this system is that it is a pain to empty. The urine container comes with a handle, but as I noted earlier, the poop chamber does not. Carrying the poop chamber against my chest or in front of me to the compost pile is not one of the most pleasant things I've ever done. It would be nicer if there was a handle.

Conclusion

If you understand the importance of rebuilding soil by returning nutrients contained in human excrement and urine, a sawdust toilet may be just what the doctor ordered. They are simple to use and inexpensive and provide an excellent way of capturing all of your urine and feces for reuse. As noted in the chapter, sawdust toilets do require a humanure compost pile to safely convert excretions into safe and valuable soil nutrients. In Chapter 5, I'll describe everything you need to know to build, establish, and maintain a humanure compost pile in your backyard so that neighbors won't have the foggiest idea what you're doing. In the next chapter, though, I'll show you ways to build your own indoor composting toilet that performs as well as, or even better than, the commercially available composting toilets I've installed and used.

How to Build Your Own Remote Composting Toilet

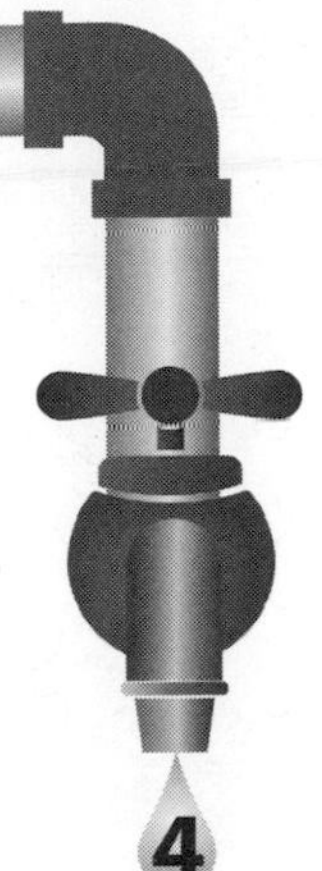

Commercial composting toilets are fairly expensive and, based on my experience and experience shared by other users, can be rather challenging to get to work correctly. Although some people I've interviewed have used dry and wet composting toilet systems successfully, I retired my Envirolet dry remote composting toilets and designed and built a composting chamber to replace it — partly because, as a do-it-yourselfer, I like to design and build lots of things. I still use my Sun-Mar, but because of issues with leakage and difficulty getting it to work right, I also use a sawdust toilet and a Nature's Head toilet for the classroom.

My homemade remote composting toilet is easy to operate, reliable, leak-proof, and extremely cost effective. It is also generally odor- and insect-free — if managed correctly. All my composting alternatives now in operation require an outdoor compost pile, a subject I'll discuss in Chapter 5.

If you can't afford to buy a commercial composting toilet or want to retire your commercial composting toilet that's not working right, you may want to consider making your own. There are lots of good ideas in books and on the Internet, and many of them work well — no matter the climate zone you live in. Most designs are for remote composting toilets — that is, toilets in which the composting chamber is located in a basement or perhaps another room (utility room) below the throne. In this chapter, I'll discuss a remote composting toilet I built and have used with great success for several years now.

As you study this material, remember that homemade composting toilets can be designed and built for outdoor use, too — for example, in outbuildings like a workshop or specially built composting outhouses. Before we go into details, let's examine some key principles of effective composting toilet design that apply to indoor and outdoor composting toilets. These are features you must incorporate for a system to work properly. Keep these in mind when you design your own system. If you do, your chances of success will be greatly improved.

Guidelines for Designing and Building a Composting Toilet

An effective composting toilet must create an environment conducive to microbial decomposition of its contents. Moreover, it must do so without leaking, smelling, and becoming an insect breeding ground — and hence a point of contention in a marriage or between family members. To make this happen, here's what you need to do.

Build It Tight — Watertight, That Is!

First and foremost, be sure the system you build is watertight so it does not leak *under any circumstance.* Over the years, both commercial dry and wet remote composting toilets have leaked on me for one reason or another, often creating significant messes. I know it won't make manufacturers happy, but in my experience, their designs had very little tolerance for error. Composting toilets should be designed never to leak under any circumstances.

For your health, safety, and peace of mind — and to ensure marital harmony — build a system that will not leak. Period. Even when improperly used. Think worst case scenario and design for it. For example, even if the receptacle fills with ten gallons of urine from a holiday party, be sure there's no chance that your composting chamber will leak.

Install a Drain

A waterproof composting chamber is a must. To meet this goal, design the system so that liquids from the composting chamber can be safely drained out of it. (I'll show you the design I've come up with to ensure this.) Another option is to divert all urine from the composting chamber — that is, at the toilet — by installing a urine diverter or a urine-diverting toilet. This design ensures that urine flows into a separate waterproof chamber that can then be emptied or run into a drain. Properly designed and installed drains remove liquids from both wet and dry composting toilets.

Fig. 4-1: Bulkhead. *These commercially available fittings are used to penetrate plastic tanks and pond linings, for example, to install plastic drain lines. They are sturdy and work extremely well.*

Be sure the drain pipe attachment to the tank is super watertight. Install a waterproof fitting called a bulkhead in the side of the tank or make your own (Figure 4-1). I'll explain how you can make your own shortly. Liquids can be piped into a nearby floor drain in clear plastic tubing. Be sure to anchor it with zip ties to the drain cover.

To reap the full benefits of your nutrient-rich deposits, it's best to collect the leachate in a container located near the compost chamber. Capturing the leachate allows you to put it to good use. As noted in the previous chapter, it can be diluted and used to fertilize plants. We pour ours undiluted directly on our pastures.

When installing the drain on a compost chamber, be sure that it is at the lowest point in the system so all urine drains out. Also be sure to install a drain that won't become

Liquids released from a composting toilet are derived primarily from urine but also from feces — after all, feces are about 75% water! The liquid that drains from a system without a urine diverter is brown. It looks a lot like coffee. The brown color is created by urine that flows through the decomposing fecal matter/cover material/toilet paper mix. More specifically, urine flowing through the decomposing fecal matter dissolves some of the feces, creating a deep brown broth — a leachate. It's amazingly odor-free.

clogged by any sludge that seeps into the liquid collection area below the composting zone. (You'll understand what this means shortly.)

Also, be sure to install a valve on the drain pipe. You'll need a valve you can shut off so when you empty the urine collection vessel, leachate doesn't continue to drip onto the floor. You'll also need it when hauling the composting chamber outdoors to your compost pile to empty it. (More on this shortly.)

Remember, too, the less water that's deposited in the system, the less likely it will leak. In my view, dry (waterless composting toilets) are your best bet. As much as I love one-pint flush toilets, I'd steer clear of them. Under heavy use, the water adds up rather quickly. If at all possible, design your system for direct deposit from a toilet above to the composting chamber below. If a dry system is not possible, for example, the tank can't be located immediately below the toilet, you will very likely need to install a one-pint flush toilet. In such cases, be sure to install a bomb-proof drain system. Also, be sure that any penetrations you create in the unit, for example, for stirring the contents to promote aeration, won't leak — under any circumstances. Seal all penetrations and whenever possible place them above the high water line.

Fig. 4-2: Garbage Bin on Wheels. *This is my favorite garbage bin for making compost chambers. Be sure to buy a sturdy one with a large, flat lid like this one. Wheels make it easy to transport composted humanure out of your basement and to your compost pile.*

In my research, I've found that there are numerous ways to build a remote composting toilet compost chamber. Plastic 55-gallon barrels can be used to create a good watertight chamber. However, I prefer sturdy household trash bins on wheels like the one shown in Figure 4-2. I used one to create my remote composting toilet composting chamber. You can also build a permanent compost chamber from cement block or concrete (Figure 4-3). Seal it well and provide a drain system. Also be sure to design a watertight access that makes emptying the chamber easy.

To seal a permanent vault, I'd recommend using environmentally friendly foundation sealants — that is, a rubber-based sealant that does not contain potentially dangerous volatile organic chemicals found in traditional petroleum-based foundation sealers.

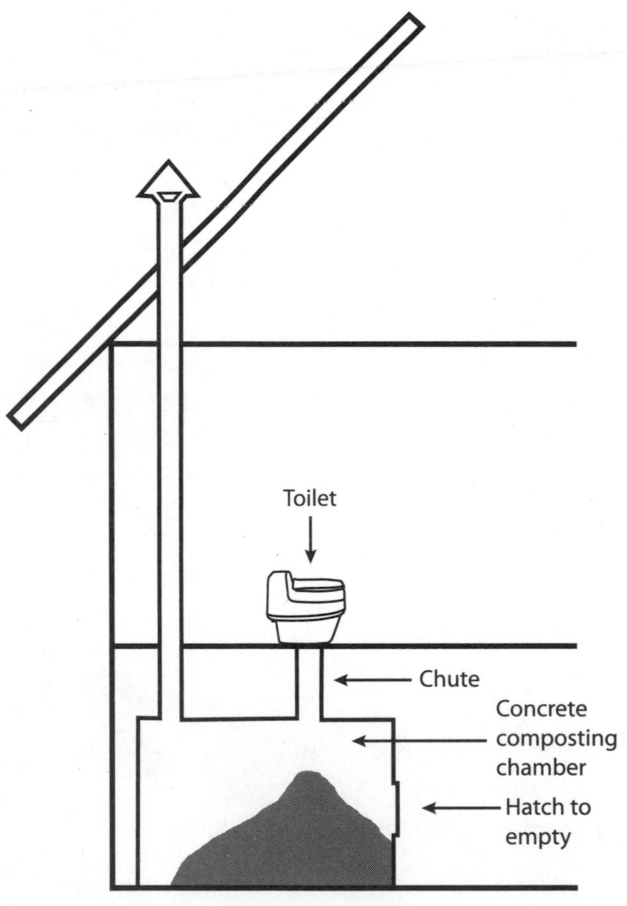

Fig. 4-3: Concrete Compost Chamber. *You can build a waterproof concrete compost chamber in your basement, but be sure it won't leak.*

Seal It Tight, Ventilate It Right!

Once you've settled on a waterproof composting chamber with drainage, the next consideration is ventilation. To keep your home smelling nice, you will need to install a vent pipe to transport odors from the composting chamber to the outdoors. A 2-to-3-inch PVC pipe should work fine.

Vent pipes typically run through interior walls, then the attic, and then they penetrate the roof. This way, any odors that escape from a system will be released away from the sensitive noses of friends and family members.

Retrofitting an existing home with a vent pipe can be challenging. If a vertical route is problematic, you may be able to run a vent pipe laterally out through a wall, then up the side of the exterior wall of the structure. The top of the vent pipe needs to extend well above the roof line. Check with your building department for requirements on proper venting.

White Schedule 40 PVC pipe is usually used for vent pipe. Unfortunately, this product is not UV resistant. So, when running pipe outdoors — for example, above the roof or side of your building — be sure to paint all exposed pipe with latex paint to protect it from UV radiation.

Also, bear in mind that rainwater can enter the vent pipe and add unwanted moisture to your composting humanure. Be sure to install a turbine or a vent pipe cap to keep rain and snow from entering your composting chamber (Figure 4-4). Some people install vent pipe caps with fine-mesh screens to prevent insects from entering their systems. You can purchase these online. You may want to try an E-cap vent pipe cap from Peak Industries (Figure 4-5).

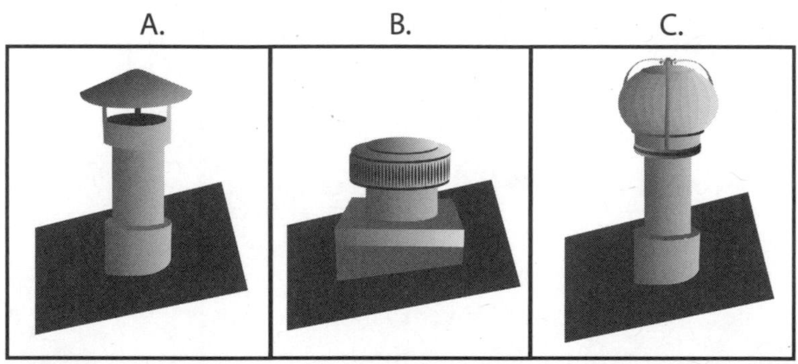

Fig. 4-4: Keep Water Out. *To prevent snow and rain from entering your compost chamber be sure to cap off the vent pipe. Here are some of the common options.*

Fig. 4-5: E-Cap Vent Pipe. *This relatively innovative design works well to prevent rainwater from entering a vent pipe.* SOURCE: PEAK INDUSTRIES

This design helps prevent wind from blowing down into your compost chamber. Cold air blowing into the system will slow the rate of decomposition. It could also create back drafting — that is, it could force odoriferous gases into your home.

Remember, too, that ventilation works best if you minimize bends in the pipe. Try for a straight run from the composting unit to the roof. Avoid 90-degree bends, if at all possible, as they will dramatically reduce air flow out the vent pipe.

To assist in ventilation, you may want to install a small, low-wattage, and quiet (low-sone) fan in the vent pipe. Fans not only help force smelly gases up the vent pipe, they help remove water from the soggy decomposing fecal matter/organic cover mix. If designed properly, a fan will also aid in aerating the composting materials, accelerating aerobic decomposition.

To save energy, you may want to install a small solar-powered DC fan on a homemade composting toilet. You can mount a small 12-volt solar module on a ground-mounted rack or pole on the south side of your house. When the sun shines on the module during the day, the fan will automatically turn on. When the sun goes down, the fan will turn off. If you want 24-hour-per-day ventilation, install a small AC fan or a DC fan with a battery to provide power at night or when the sun isn't shining.

While we're on the topic of vent pipes, be sure to insulate vent that runs through attics if you live in a temperate or cold climate. Insulating the vent pipe will reduce the amount of water that condenses inside the pipe and subsequently drips back into the compost chamber on cold winter days and nights.

While venting is important, active venting with a fan has some downsides. I discussed this topic in Chapter 2, but because it's so important to energy efficiency and sustainability of your home, it's worth repeating here.

The biggest problems with fans — and no one talks about this — is that they reduce the energy efficiency and the comfort level of a home. That's because air that is vented from inside a house via the composting toilet in the winter will inevitably be replaced by cold outside air. (Air that is vented outdoors must be replaced by incoming air.) In our homes, indoor air that is vented outdoors is replaced by outdoor air that seeps in through tiny cracks and other openings the building envelope — the walls, floors, roof, and foundation. If you are pumping 100 cubic feet of inside air out every minute, 100 cubic feet of cold outside air must enter to replace it. (A cubic foot of air is about the size of a basketball.)

In the winter, exhausting warm air from the building through a compost chamber and vent pipe and replacing it with a steady stream of cold outside air has many adverse effects. First, it increases demand for heat. It will therefore increase your energy bill and your carbon footprint. It could even make the home much less comfortable. In the summer, just the opposite occurs. The compost system vent removes cool indoor air and replaces it with warm outside air. Once again, this reduces comfort and increases energy costs.

The ideal solution to this problem would be to avoid venting indoor air and replacing it with outside air. This can be done by running a dedicated air-intake pipe from the outside of your home directly into the compost chamber. Thus, the only air that enters the composting chamber comes directly from outdoors. Indoor air temperature is not compromised by venting the composting toilet, and annual energy consumption for heating and cooling is reduced.

To make this system work in the winter, however, you should find a way to preheat the incoming air. In the winter, cold outside air entering the compost chamber could drastically lower the temperature — slowing down, perhaps halting, microbial activity entirely.

To prevent this, incoming air should be heated to room temperature before it enters the compost chamber. This can be achieved by running the pipe underground, well *below* the frost line, before it enters the building, as shown in Figure 4-6. If the run length is sufficient, the air will be naturally heated because the Earth's temperature below the frost line remains around 50° to 55°F (10 to 13°C) in temperate climates year round. (In desert

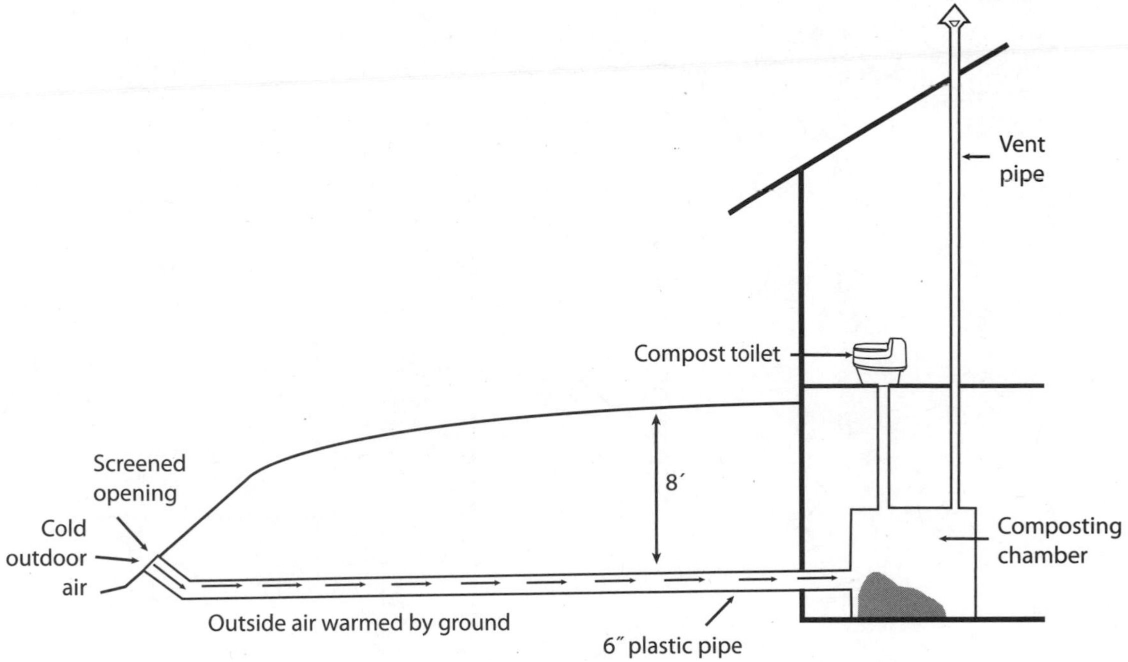

Fig. 4-6: Earth-warming Tube. *To prevent your composting toilet from sucking air out of your home and causing cold air to leak in to replace it, consider installing an earth-warming tube to heat incoming air. This will help maintain temperatures required to ensure continuous composting.*

climates, it remains around 70°F [21°C].) The pipe may need to run inside the building for some distance as well to heat it more, because 50° to 55°F (10° to 13°C) incoming air will retard microbial decomposition.

If you are building a new house, plumb this pipe into your house before you lay the foundation. I'd recommend at least a 3-inch (7.6 cm) PVC pipe, equipped with a fan, to suck air in. Let it run underground at least 20 feet (6.1 meters). Be sure that the opening to the outdoors is covered with a fine-mesh screen to keep critters out. Be sure rain and snow melt can't enter either. Seal the foundation penetration well, too. For optimal performance, place the pipe on the south or sunny side of the building.

Another option is to build a small solar hot air collector and run the air from this unit, through the wall, into the composting chamber (Figure 4-7a and b). Be sure to insulate the pipe run from the collector to the compost chamber well. Solar hot air collectors can be made from wood or metal. You'll want to install some kind of absorber plate — a dark metal surface to absorb sunlight.

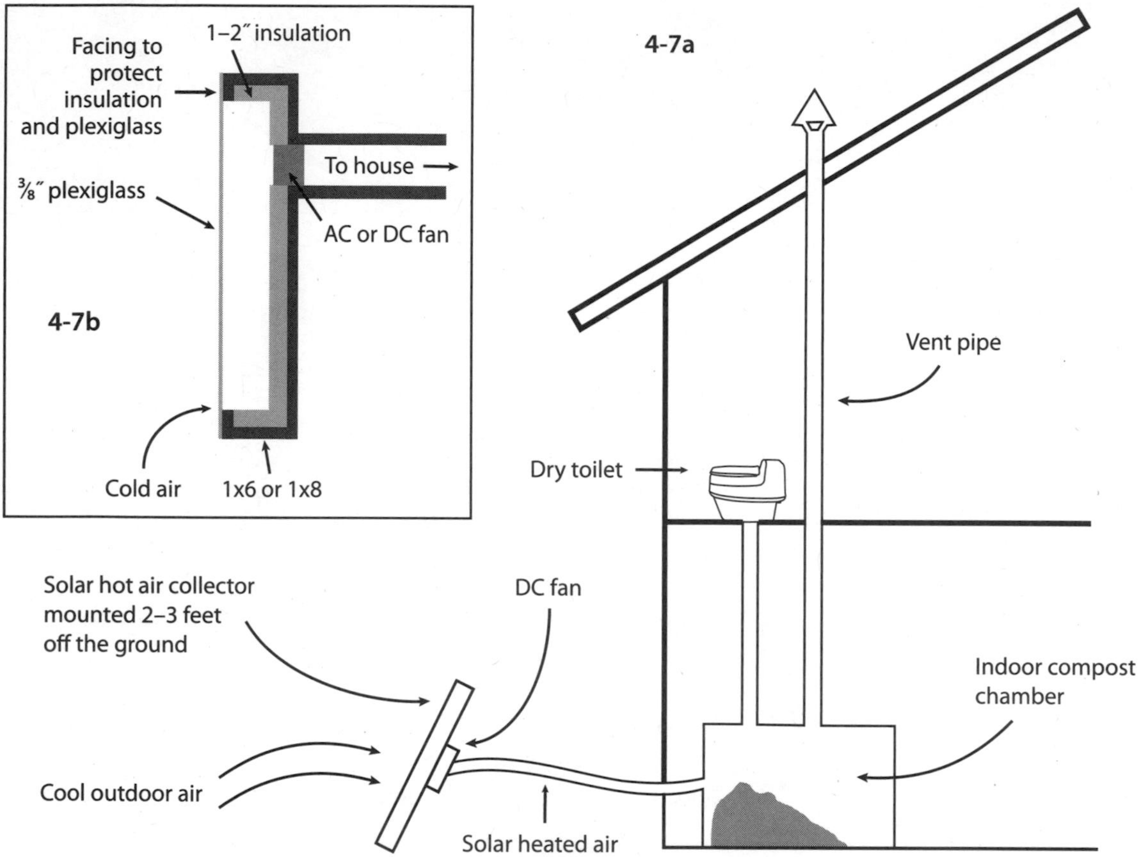

Fig. 4-7a and b: Solar Hot Air Collector. *You can build your own solar hot air collector to preheat air for a composting chamber.*

I'd recommend using metal roofing. Enclose that in an insulated wooden box with a glass or plastic cover. Mount the unit on the south side of the house.

Make It Easy to Empty

A third requirement for a successful homemade composting toilet is a convenient way of emptying the unit. If the composting chamber is fixed — say made out of cement blocks in your basement — be sure to install a door that permits easy removal of the composted human "waste." Be sure the door is placed above the water line — the level urine collects. And be sure it is tightly sealed so odors can't escape. A tight seal should also prevent water leakage, should the system flood.

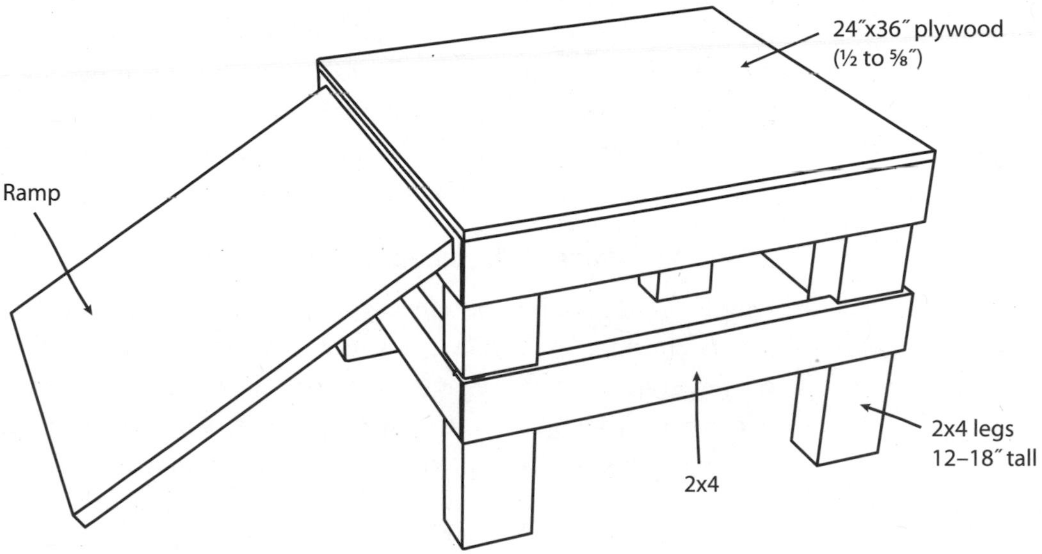

24″x36″ plywood (½ to ⅝″)
Ramp
2x4 legs 12–18″ tall
2x4

Fig. 4-8: Platform for Compost Chamber. *This is a drawing of the platform I built for my compost chamber. Raising the chamber off the ground makes it easier to drain water from the system.*

If you build a composting chamber out of a plastic garbage container, think seriously about elevating it on a small ramp or platform (Figure 4-8). That way, you can install a drain pipe near the bottom of the unit. Leachate will flow into a bucket or floor drain via gravity.

You may want to build two compost chambers, so after you remove the first one you can let the humanure/cover material age for a while longer. This allows additional decomposition of the more recently deposited materials. The second unit can be wheeled into place immediately and used as the first one continues to age.

Whatever you do, don't skimp on containers. Purchase a sturdy plastic trash can or trash bin on strong wheels. Don't use a plastic garbage can or bin that's been out in the sun, getting zapped by UV radiation. UV weakens plastic, making it brittle. You're just asking for trouble. Also

For Ease of Operation

For ease of emptying a compost chamber, consider purchasing a garbage bin that comes with sturdy wheels (Figure 4-2) or can be placed on a platform with sturdy wheels. That way, you can roll it out of the basement rather than carry it out to be emptied. A sturdy dolly would work as well.

consider the size of the container. If the bin is too large, it could become too heavy and hence could be difficult to empty. This is especially important if you are dumping the contents of the composting chamber into a compost pile for further processing. (I'll provide recommendations on the size of the chamber later, when I describe how you can build one.)

Ensure an Oxygen-Rich Composting Environment

Fourth, and of equal importance, you will need to design the system to remain aerobic. What this means is that you need to find ways to ensure that the organic mixture (feces and cover material) does not collapse on itself, eliminating air spaces. If this happens, the aerobic (oxygen-loving) bacteria will become inactive and anaerobic bacteria will take over. At that point, the unit will begin to generate methane. Methane is a combustible gas. (It's the main combustible gas in natural gas.) Anaerobic digestion also produces some pretty awful-smelling odors. So, how do you ensure that the system remains aerobic?

First, be sure to add an organic cover material that provides loft. As you may recall, loft means there are lots of air spaces in the material. This permits air (containing 20% oxygen) to enter. The more oxygen, the happier your aerobic bacteria will be.

Some people throw stale popcorn into their compost chamber, though I have no idea how someone could let popcorn go stale! Others use peat moss, a product I avoid because of environmental impacts of harvesting it. Others use sawdust. Finely chopped straw works. I gather dry leaves in the fall and crush them *and add them along with straw and sawdust (I use hydrated pine pellets rather than sawdust from a mill)*. I've even added well-aged cow manure/compost mix. It doesn't matter, so long as it is organic and fluffy. I store bags containing all these materials next to my compost chamber in the basement and alternate them — just for fun. (Who said country living is all work!?!)

Another way to help ensure aerobic conditions is to design the system to prevent it from becoming flooded and, hence, waterlogged. Excess moisture reduces oxygen content and causes the materials to collapse or settle. As just noted, wet material compresses older damp material, closing off air spaces.

Another excellent way to maintain the air spaces is to periodically agitate the mix. For example, you can install a paddle type mixer attached to a crank that helps turn the decomposing material. The agitator on the Nature's Head toilets, discussed in Chapter 3, is a good example. Be sure any agitator that is operated from outside is sealed where it penetrates the side of the composting chamber. You could mix the material periodically by hand using a sharpened stick or a metal implement. (Though this requires dedication to the cause!)

Keep It Hot!

The fifth requirement for successful composting is heat. As you know by now, microbes generate heat as they break down organic matter. If the room stays above 60°F (15.6°C), heat production within the composting mass of humanure will help maintain proper temperature and keep decomposition running smoothly. In December 2014, as I wrote the second draft of this chapter, it was 62°F (17°C) in our basement and 80°F (27°F) 6 inches (15 cm) below the surface of the composting mass in my composting chamber.

To ensure adequate internal heat temperature, you may want to insulate your composting chamber. Be sure to elevate a composting chamber off cold concrete floors, too! Cool concrete or soil beneath a concrete or plastic composting chamber draw heats out of the unit by conduction. If you doubt this, lie down on your basement floor. Even in the summer, this will cool you down considerably.

Fitting a moveable container, for example, a garbage bin on wheels, with insulation can be tricky, so you might consider framing in a small insulated room in your basement to house the composting chamber.

Keep the Bugs Out

The last thing in the world you want inside a compost chamber is insects, especially flies. Insect invasions can be prevented by ensuring that the system is airtight. Be sure lids or doors you build are

Eliminating Fruit Flies

I have had problems with fruit flies in indoor composting units, such as my sawdust and Nature's Head toilets. This occurs when I deposit the skins of citrus fruit such as oranges and grapefruit in my systems. Here's what I've found: citrus fruit skins often harbor eggs of fruit flies. The eggs hatch inside the toilet, producing swarms of flies that quickly became a nuisance. To eliminate them, I sprinkle a couple cups of diatomaceous earth on top of the mix inside the poop bin, which has always seemed to take care of the problem. Apply generously.

sealed with rubber or foam gaskets. Also, be sure that family members close the lid on the throne when they're done. In addition, cover all fecal matter at the end of every day's use with sawdust.

Some people install fine-mesh screens on their vent pipes or screened vent pipe caps, as noted earlier. To err on the cautious side, you may want to start with a screened vent pipe.

We occasionally encounter fruit flies and house flies in our compost chamber in the summer. To eliminate them, I sprinkle diatomaceous earth over the feces. Diatomaceous earth is available in 50-pound (23 kg) bags at feed stores or in smaller quantities from home improvement centers like Lowe's in the United States.

Ensure Ease of Access

The final requirement for a successful composting toilet is to be sure to install the composting chamber in a location that permits easy access. Before you drill a hole in your floor for the toilet's poop chute, set everything up to see if you can easily maintain the system and remove waste without standing on your head or crawling on your belly. Is the space easy to get to? Is there room for a shovel and buckets? Will it be it easy to remove the humus or roll the bin outside? Is the space well-lighted? Will you be able to easily install vent pipes? Is there adequate storage room for cover material/bulking agent? Is there enough room for a bucket to hold the leachate? You get the idea, no?

Building Your Own Composting Toilet

As you would expect, the Internet and books provide numerous plans for building self-contained and remote composting toilets.

The easiest self-contained "composting toilet" is the sawdust toilet, combined with an outdoor compost bin, a topic I describe in Chapters 3 and 5.

This chapter focuses on dry remote composting toilets — those in which the toilet throne is installed on a floor immediately above the composting chamber. To build one, you'll need to live in a home with a basement. You'll also need room in the basement immediately below the toilet to install an easily accessible composting chamber.

Remote dry composting toilets can also be installed in two-story homes without basements. In such instances, the throne needs to be installed in a second-story bathroom, but — and here's the catch — you'll need a room immediately below the toilet to install a composting chamber, preferably not a bedroom, your kitchen, dining room, or family room.

Fig. 4-9: Our Remote Dry Composting Toilet Throne. *This is the throne we use with our homemade composting toilet. The throne came from Envirolet and was part of their remote dry composting system that we retired for various reasons. Notice the bucket in the foreground. It contains used toilet paper that is composted outdoors in our compost bin along with deposits from our various composting toilet systems.*

If you don't have a basement or a two-story home where you can set up a remote, waterless toilet, I'd recommend installing a sawdust toilet or a commercial self-contained composting toilet.

To begin, let's start with the throne.

The Throne

The throne of a remote dry composting toilet system can be built build out of plywood, much like the cabinet of a sawdust toilet. However, homemade thrones can be tricky to build, and are often a bit rustic. Because of this, I'd recommend buying a throne designed for remote dry composting toilet systems like the one shown in Figure 4-9. Thrones can be purchased from commercial composting toilet manufacturers such as Sun-Mar and Envirolet.

As shown in Figure 4-9, commercially made thrones are fairly attractive and fit into almost any bathroom décor, unless it's a top-end bathroom with outrageously expensive fixtures. Chances are, an attractive, well-built throne will reduce the reluctance of skeptical friends, family members, and relatives. Because composting toilets are a far cry from what most people in more developed countries are used to, making the experience more palatable (probably a bad choice of words) may help win over skeptics — or at least reduce resistance.

I installed a waterless composting toilet from Envirolet in my house, but after a year of problems with leaks, yanked the compost chamber in the basement and built my own composting chamber, which is performing remarkable well. It's never leaked!

Commercially made toilet thrones will set you back about $400, at this writing. Envirolet's thrones come in a variety of colors. Add another $90 for a color. You can check out the color options on their website, although it's a good idea to view them in person, if possible. (Colors on computer screens may not match the product's color.) To view, go to Envirolet's home page, then scroll down and click on "waterless remote" then "colors." Sun-Mar's toilet costs about 20% less, but there are fewer color options.

To install a waterless throne, you'll need to cut a fairly large hole in the floor to accommodate the poop chute — usually a 6-to-8-inch PVC pipe or flexible 8-inch black tubing. You'll also need to bolt the toilet to the floor for stability. You don't want it moving around when in use.

Both Envirolet and Sun-Mar toilet thrones come with removable "bowls" (Figure 4-10). As shown in Figure 4-10b, they're not really bowls, but rather removable plastic sleeves that are open on both ends. Removable "bowls" are designed to facilitate cleaning. Without this feature, cleaning would be a nightmare.

Over the past few years, we have found that the lining of our Envirolet toilet gets pretty dirty pretty quickly. While feces generally typically drop straight down, some do ricochet off the sides, leaving unsightly skid marks. Urine drains down the sides, leaving mineral deposits. Diarrhea is especially troublesome. If someone has diarrhea, remove the bowl and rinse it out before the stuff dries. Otherwise, it's very difficult to remove. I tried to scour the liner coated with dried diarrhea with a garden hose. That failed. I then tried a power washer. That failed, too. I then soaked the liner for two days in an 18-gallon plastic tub filled with soapy dishwashing detergent containing a digestive enzyme. I then rinsed it with a high-pressure nozzle on our garden hose. No luck. Because my power washer went caput, I ended up scrubbing the liner with a rag. Even after two days of soaking in water and detergent, hand cleaning took some strenuous effort. I'm sure the power washer would have worked at this stage. It's not a job I would like to tackle again. If you don't own a power washer, or can't borrow one, try scrubbing the bowl out with a long-handled brush.

I routinely clean my bowl even without diarrheal deposits and it can be quite a challenge, even with a power washer. To make your life easier, check the lining often and clean it as soon as it becomes soiled.

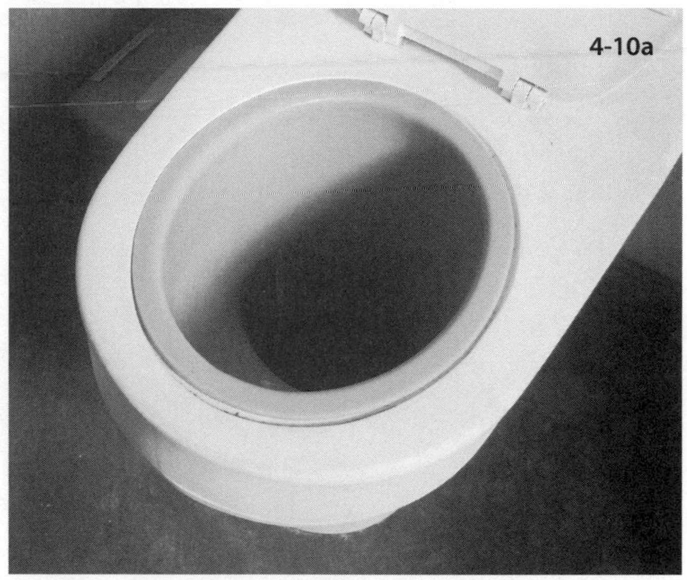

Fig. 4-10a and b: Toilet Bowl. *This nice-looking toilet (throne) made by Envirolet was used by the author in his homemade remote composting toilet. (a) Notice the removable bowl (lining) which makes cleaning much more pleasant. (b) Bowl removed.*

Designing and Building a Composting Chamber

Fig. 4-11a and b: Rubbermaid Brute and Dolly. *(a) This garbage can and (b) dolly make an excellent compost bin. They are extremely durable and easy to work with. The Brute has a very flat lid, which easily accommodates the vent pipe and poop chute.*

Once you've tackled the throne, you'll need to design, build, and install a composting chamber — one that will be placed one story below the throne smack dab in the middle of the drop zone. As noted earlier in the chapter, there are a lot of options when it comes to creating a sturdy, leak-proof composting chamber. If you want to try a simple and inexpensive method, I'd recommend either a plastic trash can or a trash bin. Let's start with trash cans.

For best results, start with a heavy-duty plastic trash can with a sturdy lid and handles — preferably one that's on wheels. Rubbermaid sells a heavy-duty 44-gallon commercial trash can, appropriately named the "Brute." It cost me around $50 in 2014. The Brute has thick walls and extremely sturdy handles (Figure 4-11a). Another desirable feature of the Brute is its large, relatively flat lid. This makes it easy to cut holes for the vent pipe and poop chute.

Rubbermaid also sells a dolly custom made for Brute (Figure 4-11b). The dolly is a round platform on which the Brute sits. It comes with five fairly heavy-duty casters (wheels) that allow you to roll your new compost bin easily on a concrete basement floor. The dolly/platform costs another $35, but is well worth it.

Another option for a composting chamber, one I am currently using with great success, is a wheeled trash bin like the one shown in Figure 4-2. Trash bins run from 32 to 96 gallons. I chose a 64-gallon Toter (indoor/outdoor trash garage can). It is well made, sturdy, and was manufactured from 30% recycled plastic. I purchased it at Lowe's for around $75 in 2014.

I wouldn't recommend 48-gallon bins. They're much too small. Nor would I recommend Toter's (or other manufacturers') very large 96-gallon bins. Once full, they are very heavy and difficult to empty.

I'd recommend starting with a 64-gallon bin, and see how you like it. If, after using it for a year or two, you think you can handle a larger unit, go for it! The large 96-gallon units cost between $90 and $100 in 2014, depending on the manufacturer.

Wheeled trash bins can be purchased at home improvement stores such as Home Depot, Lowe's, or Menards throughout North America. I've noticed that these stores often offer a much greater selection online than you'll find in the stores, so you may consider ordering one online. When I checked, Home Depot seemed to have the best selection. The only problem with ordering online is that you won't get to study the bin beforehand to see if it is going to work. However, you can order a unit, pick it up at the store, and then return it if it's not quite right. They'll understand.

Urine Drain

To drain the urine from my compost chamber, I installed a ½-inch plastic PVC drain 2.5 inches from the bottom of my bin. On the second bin I built, I placed the drain at the very bottom so no urine collects in the bottom of the composting chamber.

I'd recommend drilling the hole for the drain pipe on the *side* of the bin. That way, the drain will be out of the way when you are wheeling the unit to the compost pile. If located on the back of the unit, it could catch on obstructions when wheeled to the compost pile. If located on the front of the unit, it could snap off when you are emptying the contents of the bin into the compost pile.

Fig. 4-12: Compost Chamber Drain and Homemade Bulkhead Details. *Photo shows the details of drain with valve and homemade bulkhead I installed on my compost chamber.*

To create a clean hole for the drain, I'd recommend using a spade bit, hole saw, or a Forstner bit. Forstner and spade bits are easiest.

To create a watertight seal for the drain pipe, you'll need a watertight fitting in the hole you drilled in the side of the bin. One option is to purchase ½-inch bulkheads, like the one shown in Figure 4-1. Bulkheads are watertight fittings. They can be ordered online and may also be available at farm supply stores. (Farmers use bulkhead fittings for tanks, but tend to install larger ones.)

Another option is to make your own fitting. This is what I often do. Details of a homemade bulkheads are shown in Figure 4-12. As illustrated, bulkheads can be made from a ½-inch PVC male adaptor and ½-inch female adaptor. Both male and female adaptors are threaded on one side and unthreaded on the other.

A rubber O ring is installed to create a watertight seal. The male adaptor screws into the female adaptor, and the O ring is placed between them (on the inside of tank). I use No. 15 rubber O rings (No. 15 O rings have a 1-inch outside diameter [OD] and a ¾-inch inside diameter. The ring is ⅛-inch material.)

To install, I insert the male adaptor through the tank from the inside with the O ring already installed on the threaded portion. While you hold onto the male adaptor, a helper can screw the female adaptor on from the outside of the bin. Be sure it is tight.

Also be sure that hole you drill is not too big. It should be just large enough to allow the threaded end of the male adaptor to pass through. If the hole is too large, it will leak.

To drill holes in my bins for the ½-inch PVC male adaptors, I used a ¾-inch Forstner bit. The ⅝-inch bit is a bit too small, pun intended. I'd recommend testing bit size on a spare piece of plastic first to be sure the bit size you choose ensures a good, tight fit for the male adaptor.

Once the homemade bulkhead was in place, I slipped a 6-inch length of ½-inch PVC pipe into the unthreaded end of the female adaptor so it extends over the edge of the platform on which I placed my compost chamber. I then attached ½-inch PVC valve onto the pipe, and slipped another short section of ½-inch pipe into the free end of the valve, as shown in Figure 4-12. To that I attached the ½-inch female fitting. This fitting is threaded internally on the free end (hence the designation "female"). Into that, I screwed a threaded nylon barb fitting, also shown in the photograph. The barbed end was sized for ½-inch clear plastic tubing that I attached and ran into a 1-gallon plastic jug to collect the urine leachate. Initially, I placed the 1-gallon jug inside a 5-gallon plastic bucket that served to catch overflow — for example, if I neglected to empty the 1-gallon bottle in time. I found that the 1-gallon jug filled up pretty quickly. To reduce maintenance, I now just let the leachate collect in the 5-gallon plastic bucket. I empty it once a week.

As shown in Figure 4-13, I installed a plastic grate inside the garbage bin to separate solids from liquids. I used a 2-foot x 4-foot

Fig. 4-13: Plastic Grate. *This grate was once used for fluorescent lights. It makes an excellent grate for homemade compost bins.*

plastic grate from a fluorescent light fixture. Because plastic grates like these are no longer in fashion, you may find it difficult to locate one. I purchased mine at a small home improvement center in a town about 20 miles from my residence. I couldn't find one at Lowe's. If you can't find one locally, you may be able to locate one on the Internet. The grates cost me about $12 each, and each grate is large enough for two compost bins.

Using tin snips, I cut a 17-inch x 17.5-inch piece out of the grate to fit inside the compost chamber. Snip one square at a time. If you are impatient, you could probably use a band saw, but you must cut carefully. A sabre saw vibrates too violently to get a clean cut. You will also need to round two edges to fit inside the slightly rounded edges of the trash bin (Figure 4-14).

To support the grate, I made a "platform" out of two 6-inch PVC pipe (Schedule 40). I cut two 10¼-inch (26 cm) pieces of

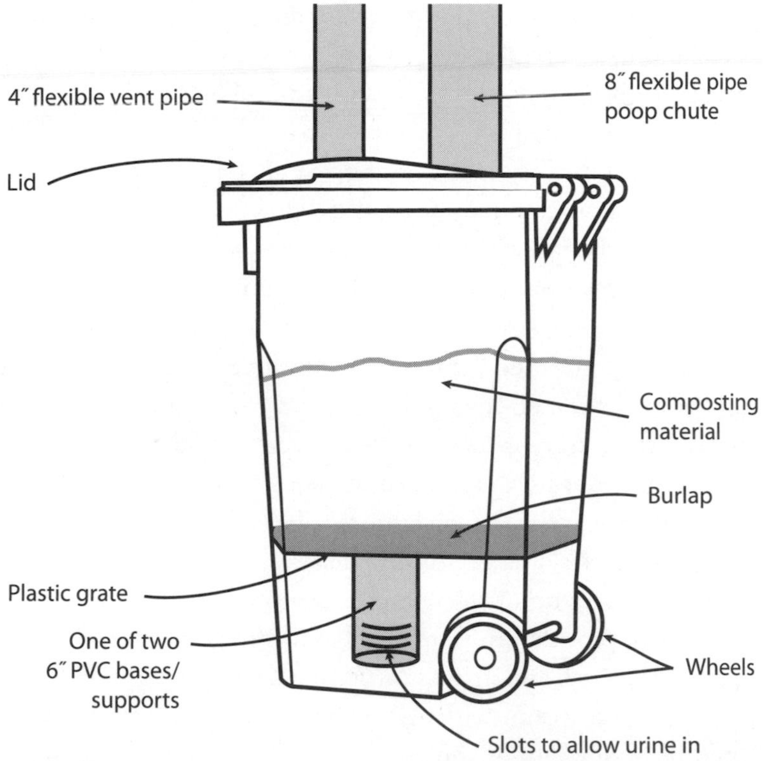

Fig. 4-14: Details of My Homemade Composting Chamber.

PVC pipe, then cut parallel drain slots along the bottom of each pipe, extending 1 inch (2.54 cm) up the side. The slots were added so that if urine enters the pipe from above, it can drain out. I used a miter saw to cut the horizontal drains lines, as shown in Figure 4-15.

I placed the PVC platform in the bottom of the bin, then laid the grate on top. The grate also balances on a ledge inside the bin that's formed by the wheel wells. In so doing, I created a 17-inch (43.2 cm) wide by 17.5-inch (44.4 cm) long

Fig. 4-15: Plastic Base. *This photo shows the simple base I made from 6-inch (15 cm) Schedule 40 PVC to help support the grate in my compost chamber.*

by 10¼-inch (26 cm) deep urine containment zone. I then placed a small piece of burlap over the grate to prevent feces and bulking agent (additive) from falling through and mixing with the urine and clogging the drain.

I connected the throne to the compost chamber by attaching an 8-inch flexible black pipe — one that I purchased from Envirolet for use with their dry toilet. You might be able to purchase this item directly from the company or find a substitute at a plumbing supply store.

To create an opening for the 8-inch flexible plastic pipe for the poop chute, I cut an 8-inch hole in the lid of the garbage bin with a jigsaw. Trace a line and work slowly and carefully so as not to deviate from your line. (I marked the cut line with a silver felt-tip pen because the lid of the trash bin was black.) I then cut a 4-inch hole for the flexible plastic tubing that attached to my vent pipe. It runs from the top of the compost chamber to the rigid 4-inch (10 cm) vent pipe that runs up through the floor and then through the bathroom wall and into the attic, and finally through the roof. (The flex pipe also came with my Envirolet toilet kit.) Flexible pipes should fit snugly into the holes to prevent odors from escaping.

Cut both holes so the pipes fit snugly. Doing so will reduce chances of odors escaping. Figure 4-14 shows all the details.

Platform

To make it easy to drain the urine from compost chamber, be sure to elevate the trash bin. I used an 18-inch high wooden platform I had retired from another application (Figure 4-8). (I had used the stand for a backup gas generator in my off-grid home in Evergreen, Colorado.) The top of the stand is plywood and measures 26 x 36 inches. A smaller, shorter platform might work better. To wheel the bin off the cart, I built a sturdy detachable wooden ramp out of oriented strand board (OSB) reinforced with 2 x 4s.

Connecting the Pipes

With the compost bin in place, I slipped the flexible 8-inch (20 cm) pipe from the throne into the large hole in the lid of my newly created indoor compost bin. I then connected the 4-inch (10 cm) PVC (Schedule 40) vent pipe to the smaller hole by a section of 4-inch (10 cm) flexible black pipe from Envirolet. Much to my surprise, they fit snugly and we have experienced very few problems with odor — even without a vent fan!

Outdoor Composting Toilets

Outdoor composting toilet systems must follow the same rules as indoor composting toilet installations. If designed correctly, outdoor composting toilets work extremely well at summer camps or for cabins, cottages, and parks — locations where they're primarily used in warmer months. They also work well year round in parks and outdoor facilities in relatively warm climates, like Florida.

As a rule, the warmer the climate, the more useful an outdoor composting toilet becomes. (That's one of several reasons why there's been so much effort to introduce composting toilets into rural areas in less developed countries.)

If you install an outdoor composting facility at a summer residence, cabin, or camp, pay special attention to preventing flies and other insects from gaining access to the composting chamber. They can be a real pain in the butt, especially if you live on a farm with livestock, poultry, or waterfowl.

Although outdoor composting toilets work best in warmer climates, they can be built and successfully operated in temperate climates with moderately to severely cold winters. When cold weather strikes and temperature inside the composting chamber drops below 55°F (12.8°C), microbial decomposition grinds to a halt. The microbes don't die, but they become dormant and temporarily stop digesting organic matter. They can even freeze without harm. When winter morphs into spring, the feces thaw, and when the temperature inside the chamber climbs above 55°F (12.8°C) the microbes inside the composting humanure resume work.

Even when the contents of an outdoor composting toilet freeze, you can continue to use it. Additional deposits merely freeze and accumulate in the composting chamber. Very little, if any, decomposition will occur until the temperatures climb once again in the spring. The cool thing about this is, in the dead of winter, you won't have to worry about flies and odors should be minimal.

To increase the usefulness of an outdoor composting toilet, you might want to consider designing a system with a well-insulated room up top to house the throne, and a well-insulated solar-heated chamber to house the composting chamber (Figure 2-11).

The easiest, least expensive, and most environmentally sound way to heat this room with solar energy is to design it for passive solar heat gain. That requires a composting chamber room with south-facing glass or durable, sunlight-resistant plastic, like polycarbonate. South-facing glass allows the low-angled winter sunlight to enter the composting chamber room. Sunlight is absorbed by the walls of room and the walls of the compost chambers and converted to heat. If designed correctly, temperatures should remain well above freezing — even in the dead of winter.

The compost chamber room (vault) wall below the throne can be made from concrete, rock, brick, cement block, or rammed earth tires. The wall provides thermal mass that absorbs and stores sunlight energy during the day that streams in through the south-facing glass. At night, the mass releases its stored heat, helping to maintain temperature inside the composting chamber.

To help maintain temperature in the winter in cold climates, be sure to insulate the outside of the chamber with rigid foam insulation. Four or five inches (10 to 12.7 cm) would be best. I'd shoot for an R-25 insulation. For optimum performance, insulate the glass at night with rigid foam.

Some people catch their excretions in plastic bins or plastic barrels. Plastic doesn't generally hold up very well to sunlight, but bear in mind that glass or plastic should block most of the UV radiation. To provide a second layer of protection, paint your plastic composting chambers with latex paint.

Some people collect their excrement in 55-gallon steel drums. I don't recommend this. Steel rusts when exposed to air and water. Moreover, steel doesn't hold up very well against acidic excretions of animals. If you use steel drums, divert the urine into a separate plastic tank to reduce corrosion. Separating urine may also help reduce odors. To divert urine, consider installing a waterless urinal for males. A urine diverter installed in the throne would work fine for ladies or men sitting down on the job.

For ease of use and to prevent flies and odors from becoming a nuisance, I'd recommend building compost chambers out of plastic wheeled garbage bins — like the one I discussed in the previous section. If the lid is pretty tight — and it is sealed with weather stripping — it is less likely to be invaded by pesky flies than an open receptacle, for example, an open 55-gallon drum or barrel.

Managing a Self-Built System

A homemade composting bin is placed into service the same way a commercial composting toilet is, a topic explained in Chapter 2. To initiate service, I sprinkle a 1-inch (2.5 cm) layer of sawdust and coconut coir over the burlap that rests on the plastic grate. I also add straw (pieces should be 1 to 4 inches long [2.5 to 10 cm]) and dry leaves. These materials are added to create loft.

You can add straw or leaves directly or shred them. To shred them, try filling a plastic garbage bin about ¼ full. Shred the straw or leaves with an electric weed whacker. Don't cut the straw and leaves up too small, however. You want the bulking agent to consist of pieces big enough to create airspaces.

Once the base layer is in place and spread fairly evenly over the burlap covering the grate, I install the poop chute and vent pipe. It is now time to begin using your under-$100 composting chamber.

To kick start the composting process, you may want to add some starter — for example, bacterial mixes purchased from garden supply outlets or from composting toilet manufacturers. Bacteria seed the system. These bacteria release enzymes that digest organic matter externally, breaking it down into foodstuffs they need to survive and reproduce. What's left over is the humus.

When I start a new indoor composting bin, I often add some accelerator that came with my Envirolet toilet. I mix it with warm water and then pour it in through the toilet, although I sometimes sprinkle it over the humanure/bulking agent after a few deposits have been made.

You don't have to buy expensive bacterial accelerators to establish robust composting. For example, I sometimes sprinkle a small shovelful of well-aged cow manure evenly over the base material as a starter. (The stuff is so well-aged that it can be applied without gloves, but then, I'm not very squeamish when it comes to such things.) Composted manure "starter" contains billions of microorganisms that start the composting process system. You could even throw in a few handfuls of top soil. It's teeming with microbes.

One of the cool features of this bin design is that the lid can easily be raised with the flexible hoses intact to inspect the composting mass of human excretion or to sprinkle cover material/ bulking agent over the poo. I add a variety of cover materials/ bulking agents, including sawdust, coconut coir, straw, dry leaves, and wood shavings. You could even add dried grass clippings — so long as they are from lawns not sprayed with herbicides or insecticides.

I add a little organic cover material to the compost chamber every day — or every other day — through the lid of the compost chamber. I could sprinkle it in from the throne up above, but the former method permits me to cover the fecal matter very well. Doing so helps eliminate odors — just as it does in a sawdust toilet. My diligence has proved to be quite effective in controlling odors. Even without an exhaust fan on the vent pipe, the system operates year round with little, if any, odor.

Over time, you'll find that human excrement, toilet paper, and cover material/bulking agent will form a volcano-shaped pile in the drop zone. To prevent this, I periodically flatten the material with a metal rod or a stick. I perform this function every month or two and mix the material to promote aeration and thus help maintain aerobic conditions.

To empty the compost bin, we roll it down the ramp, and then out the basement door to our backyard compost pile. We lift the bin onto the edge of the compost bin and flip it over. It takes two of us to lift the compost chamber. Its contents quickly empty onto humanure from our sawdust toilet and Nature's Head toilet and other compostable items.

When it's empty, I turn the bin upright, rinse it down and dump the rinse water onto the compost pile. No sense in getting it too clean, however, as it is just going to be re-filled with shit.

We typically "age" the material for another year in our compost pile before it is applied to gardens. One reason is that, if we haven't let the bin sit for six months after being removed and replaced by a new bin, the most recently deposited stuff won't be very well composted. Additional time is required to destroy any potential harmful microbes. We also age humanure in the compost pile because we like that added safety of having very well-aged, and hence thoroughly safe, humanure compost to dig into our vegetable gardens. We can vitalize our soil with very little, if any, concern for our health.

Conclusion

In this chapter, I've shown you one way to make a waterless composting toilet. There are others, so I encourage you to explore your options. Follow the guidelines I've presented in this chapter and carefully manage the system and you should be happy with your homemade composting toilet. As you know, these systems, like sawdust toilets, require a well-maintained humanure compost pile, which is the subject of the next chapter.

Guidelines for Building a Composting Toilet

- Build it watertight
- Install a drain
- Make it airtight and ventilate
- Ensure ease of emptying
- Ensure an oxygen-rich composting environment
- Keep it warm!
- Keep the bugs out
- Ensure ease of access

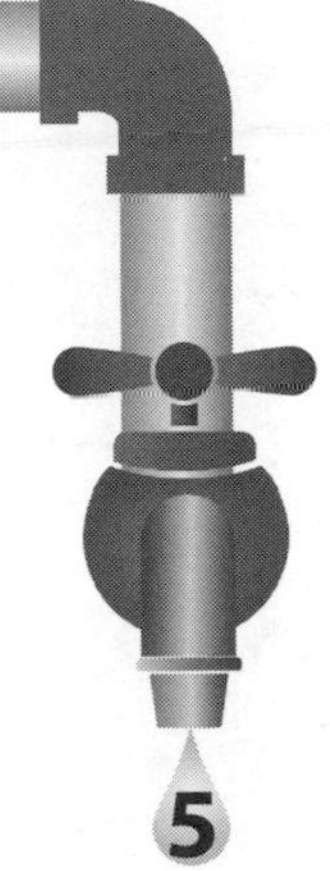

Composting Humanure — Safely and Odorlessly

To safely compost excretions collected in a sawdust toilet or a commercial or homemade composting toilet, you'll need an outdoor compost pile. As noted in previous chapters, it is best to contain the material in a compost bin equipped with at least two, preferably three, compartments. You can purchase compost bins, or you can make your own. As is often the case, you'll find many nifty designs for the do-it-yourselfer on the Internet and in many of the books on composting. In this chapter, I'll discuss how a humanure compost pile is established and maintained and show you how to build a sturdy three-chamber compost bin out of used pallets. Let's start with the construction of a bin.

Fig. 5-1: Dan's Compost Bin. *This three-chamber bin was made from used pallets in a couple of hours.*

Build Your Own Compost Bin from Pallets

Figure 5-1 is a photo of the compost bin I built to handle compostable materials from our gardens, kitchen, livestock operations, poultry coops, and composting toilets. This three-compartment compost bin was made out of pallets donated and delivered by my local lumber yard.

As illustrated in Figure 5-2, I used a 4-foot x 12-foot pallet for the back

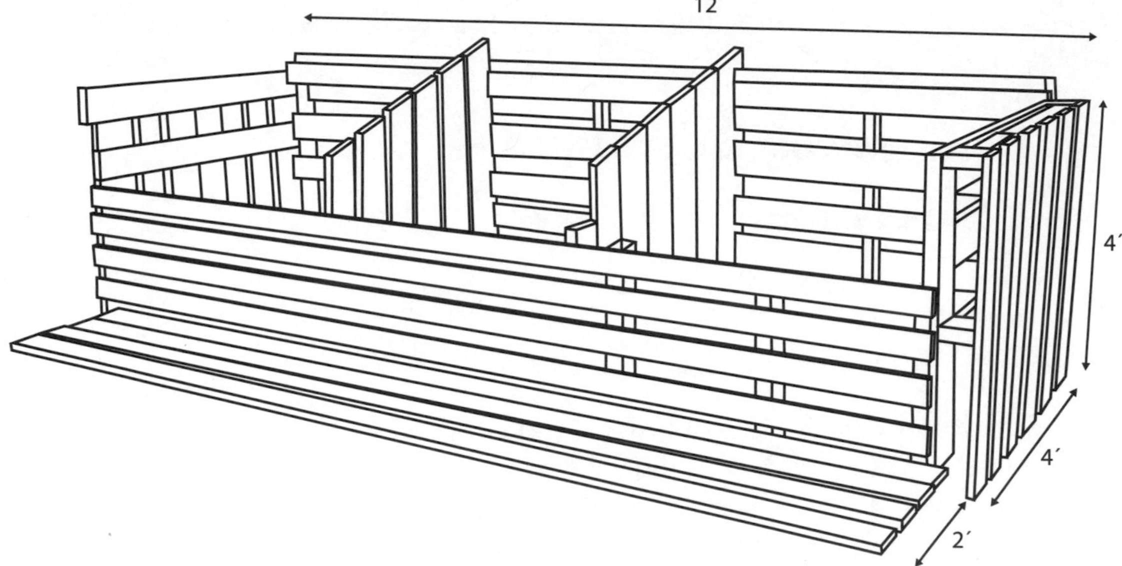

Fig. 5-2: Details of Dan's Compost Bin.

and two 4-foot x 4-foot pallets for the sides. (In metric units, the large pallet is 1.2 m x 3.6 m and the smaller pallet is 1.2 x 1.2 m). For durability and ease of disassembly, I screwed rather than nailed the pieces together. To close off the bin in front, I used a 2-foot by 12-foot pallet, attaching it at both ends. (In metric units that pallet would be 0.6 m x 3.6 m.) I then divided the massive 4-foot x 12-foot bin into three equal sections using some 2 x 4s and some reclaimed 1 x 6 pressure-treated decking. I used 2 x 4s to frame the partition walls between each bin and the 1 x 6 pressure-treated decking to build the walls.

The result of my labor was three 4 foot-long by 4-foot wide and 4 to 5-foot high chambers. (In metric units, that's 1.2 m x 1.2 m x 1.5 m.) I placed another 2-foot x 12-foot pallet in front of the bin to create a dry platform to stand on.

Next to the compost pile, I deposited a pile of topsoil I had left over from when I dug a small duck pond in my back yard. I mixed it with well-aged cow manure from our farm. Alongside this "dirt" pile, I stockpile soiled straw from the chicken and duck coops. I use the topsoil/manure and straw to cover deposits in the active pile (bin 1). This blocks odors, prevents insects from becoming a problem, and helps speed up the aerobic decomposition of the organic materials in the compost pile. It only takes a half a minute to make a deposit and cover it.

Covering a Humanure Compost Pile

Protecting a humanure pile from animals is extremely important, especially dogs — either your own dogs or your neighbors'. Our close canine friends will invade a pile and dig down to find the goodies — no matter how much cover material you place over it — if you don't install a heavy fence-wire lid.

Although I admire and love dogs, I have to admit that their dietary choices leave much to be desired. Human poop is right up there on top of the list of preferred menu items! Our guard dogs (Great Pyrenees) and our herding dog (border collie) would eat the stuff as fast as we buried it, if given the chance (Figure 5-3). It makes me wonder why we waste so much money on gourmet dog foods. Truth is, our beloved dogs would just as soon dine on fresh cow manure, cookies deposited by Fluffy in the kitty litter box, or humanure.

Fig. 5-3: Dogs, You Gotta Love 'em. *But never forget that human poo is a delicacy. Protect your compost bin from them.*

To prevent our chickens and dogs from rooting through the humanure compost pile, which also contains delectable, rotting kitchen scraps, I initially placed a 4-foot by 4-foot piece of chicken wire over the pile (1.2 m x 1.2 m) Because the chickens kept pecking in the compost pile through the chicken wire, I replaced that with a quarter-inch hardware cloth attached to a sturdy 2 x 4 wood frame. The framed screen raised the hardware cloth off the top of the compost pile, making it extremely difficult for chickens to get to the delectable treats buried in our pile. It also helped keep our dogs from scavenging in the compost pile and spreading toilet paper all over our yard. (They would sometimes pull the chicken wire off.) As an added benefit, a framed screened cover is much easier to handle than a loose piece of chicken wire.

A screen or chicken wire "lid" allows rainwater and snow melt to soak into the pile. As you probably know by now, moisture is essential for bacterial decomposition. So, cover your pile, but don't install a solid cover.

Starting and Maintaining a Humanure Compost Pile

Compost piles have received a lot of attention — in fact, entire books have been written about them. I've been composting since the 1970s and can't understand the need for a book on composting. (Sorry, to their authors.) The requirements of a compost pile are so simple that it doesn't require a lot of knowledge. Mother Nature does most of the work — so long as the pile remains moist, oxygenated, and stocked with delectable organic materials.

In many regions, leaving a compost pile exposed to the elements will help keep it moist. However, in dry regions or during extremely long, dry periods, homeowners may need to shower their piles with water from a hose every once in a while. Failing to do so will considerably slow down decomposition, perhaps even causing it to grind to a halt. But don't worry; as soon as water is reintroduced, the microbes will get back to work.

If you install a composting toilet system that diverts urine from feces or if you collect leachate from a compost chamber and periodically apply that liquid to your pile, that should suffice. If not, you may want to pour dishwater (rinse water, not soapy water) on the pile. You can also catch clean water in a bucket from your shower (as the water heats up). You could also collect rainwater to periodically moisten the pile. If these ideas involve more work than you'd like, an occasional sprinkle with the garden hose should suffice.

When starting a compost pile, you'll need a reliable supply of cover material. Just as with composting toilets and sawdust toilets, many options are available: straw, hay, grass clippings, leaves, sawdust, wood shavings, mixtures of topsoil and well-aged manure, and even weeds from the garden.

When Joe Jenkins, author of *The Humanure Handbook*, starts a humanure compost pile, he places 18 inches (45 cm) of dry organic material, such as straw, in the bottom of the bin. The straw base helps to soak up moisture from newly deposited materials from sawdust toilets. I start with a 6 to 10 inch (15 cm to 25 cm) layer of loose material, usually soiled straw from our chicken coops. I then pile the contents of our buckets from our sawdust toilet and materials from our composting toilets. I add kitchen scraps, used paper napkins and paper towels, and soiled pine pellet kitty litter — whatever needs composting. Being an equal opportunity composter, I also use the pile to process gnarly weeds, leaves, and muck I dredge from the duck ponds I clean out every year. From time to time, I add shredded cardboard. (Never throw huge chunks of cardboard in at once. It takes longer to break down.)

Compost piles can be used to process a lot more than most people think. The microbes in them will "devour" pretty much any organic material you place in this magical microbial realm. I even compost ratty old T-shirts, beat-up jeans, and other worn-out clothing. Any cloth that has outlived its usefulness is fair game, so long as it's made from natural fibers such as wool or cotton.

Composting Blue Jeans

I have been composting torn and tattered jeans for many years. After a year, all I find as evidence of my actions is the zipper. Underwear composts, too, though the elastic bands won't break down.

Contrary to popular belief, you can compost meat, dead animals, fats, and numerous other things that "experts" on composting adamantly warn against burying in a compost pile. Bottom line: if it is organic, it will break down over time. If you cover your pile well after depositing these forbidden materials and keep a screened lid on the pile, all these taboo materials will decompose and add organic matter to the rich organic humus you are generating to enrich your garden soils. (For more on this topic, I'd highly recommend you read Joe Jenkins' book.)

To compost successfully, the secret is diversity — add a diversity of organic materials, some rich in carbon and some rich in nitrogen. Most dried cover materials such as straw, leaves, wood shavings, and grass clippings add organic matter rich in carbon. This combines with the nitrogen-rich human excretions to produce an amazing organic soil supplement.

Just as in a composting toilet, cover material also helps to eliminate odors and soak up moisture. It even helps keep flies from laying eggs in the pile — and it performs these duties amazingly well. Even though we raise grass-fed beef cattle on our farm and have a large population of flies in the summer, covering the compost pile after each deposit with a layer of soiled straw or similar material does an excellent job of keeping the flies from congregating around the compost pile.

Throw a tarp over your pile of loose straw if you live in a wet climate, so it doesn't rot before you can use it. (Straw decomposes pretty quickly when perpetually kept wet.) If you live in a snowy area, cover the straw in the winter for ease of access.

Some folks use their third bin in a three-bin system to store their cover material. I just pile mine next to my "compost condo" because we're not crammed for space on our farm (Figure 5-4).

As soon as you deposit the base layer of organic matter in bin 1, it is time to start adding material from your sawdust and/or composting toilets. After dumping a bucket from my sawdust toilet on the pile, I rinse the bucket out with a garden hose, being careful not to splash. I wash down the sides with a gentle stream, and then pour the rinse water onto the compost pile. After

Fig. 5-4: Cover Material. *We keep a pile of soiled straw from our chicken and duck coops next to our compost bin year round. It makes a great cover material/bulking agent. Chickens, however, do love to scratch through loose straw and spread it around, so take precautions to prevent this.*

a couple rinses, the bucket's generally ready to be put back into service.

If you don't have a hose near your compost bins, tote your buckets from your sawdust toilets to an outdoor faucet. Add water, then cart the buckets back to the compost pile. You could also dump the rinse water in a well-mulched flower garden or a mulch bed around trees. Be careful with this water. It will contain fecal microorganisms, some of which could be pathogenic.

If you dump compost from a commercial composting toilet in your compost bin, you can follow the same directions. Remember, the buckets and trays from sawdust toilets and compost bins don't need to be spotless and sterile before reuse. They're just going to get dirty again pretty fast.

Compost piles — like humanure and cover material in a compost chamber — tend to grow like a mountain. To avoid this problem, I dump contents of sawdust and composting toilets fairly evenly across the top of the pile, spreading it out with a pitch fork. You can also divide the top of the compost pile into quadrants. Dump materials in one quadrant at a time so the pile ends up looking more like a mesa than a steep-sided volcano.

Composting Humanure Safely

Most people suffer from *fecophobia* that ranges from a mild aversion to complete and utter revulsion. To the latter, the idea of collecting excrement and urine and aging it in a backyard compost bin, then adding the end-product — the humus — to soil sends up huge red flags. I'm assuming that by now wildly fecophobic readers have long since departed our company, and listed my book for sale on Craigslist or Ebay.

For those who are solidly behind the idea, yet still mildly to moderately fecophobic — or at least have a healthy concern for what they're about to embark on — let's take a serious and sober look at safety and how one can compost human feces and urine in ways that eliminate the potential for the transmission of infectious disease.

The first thing to remember is that, although bacteria and other microorganisms are abundantly present in human feces and the world we inhabit, the vast majority of these microbes are harmless. Put another way, only a tiny percentage of microorganisms we encounter in our daily lives cause disease. Just because there are billions of bacteria living on your skin, your countertops, in the air we breathe, or in the feces we produce, doesn't mean they'll harm you — or anyone else. In fact, humans have lived in harmony with most bacteria and viruses for centuries.

The first thing to remember is that, although bacteria and other microorganisms are abundantly present in human feces and the world we inhabit, the vast majority of these microbes are harmless.

Because most microorganisms are harmless, concern over catching a communicable disease from a humanure composting system is unwarranted — under most circumstances. If you and your family members are healthy and haven't been traveling to countries where Ebola, cholera, yellow fever, and other infectious diseases exist, the microorganisms that cause these diseases should not be in your feces. Kind of logical, eh?

Moreover, as I learned early on in my studies of microbiology, most microorganisms that share our world are beneficial. As noted in Chapter 1, the bacteria that reside in the human intestinal track comprise 30% of the dry weight of our feces. Far from dangerous, most are beneficial to us. They are our allies. Our lives depend on them. For instance, some of our intestinal bacteria produce vitamins, such as B-12 and K. These vitamins are often missing from the nutrient-poor diet consumed by most people in more developed countries. Bacterially produced vitamins are absorbed into the bloodstream in the large intestine and are extremely important for our health. Vitamin B-12, for example, is involved in the metabolism of genetic material, RNA and DNA. You can't get more important than that! Vitamin K aids in the production of a number of proteins in certain cells of the body — notably, proteins that assist in blood clotting. Without blood clotting, we'd die from internal bleeding as we bump and stumble through life.

Get this: the microorganisms that live inside the human gastrointestinal (GI) tract are so vital to the proper functioning of the body that they are sometimes considered an additional organ system, much like cardiovascular system or respiratory system. To put it mildly, we can't live without them.

By various estimates, between 500–1,000 species of bacteria take up residence in the human GI tract. And — as long as we're uncovering important facts — there are ten times more bacteria in our bodies than body cells!

Over time, most people have equated bacteria and viruses — all bacteria and viruses no matter what their purpose — with disease. As a result, each year North Americans spend a fortune on antibiotics and a wide range of anti-bacterial soaps and cleaning agents to kill what are largely harmless viruses and bacteria.

Microorganisms in human feces also include numerous single-celled fungi, protozoa, and various parasites, such as roundworms.

As is the case with viruses and bacteria, most single-celled fungi and protozoa are harmless. We humans live in harmony with most of them. That said, there are some pathogenic organisms among their ranks. Malaria, for instance, is caused by a harmful protozoan. If these pathogens are in feces, they present a danger. If not, there's very little need to feel fear.

Parasites or eggs from tapeworms, round worms, pin worms, and flat worms may sometimes be present in human feces. Although they are not truly microorganisms, they are classified as such. In more developed countries, internal parasites are generally not a public health problem, except in extremely poor rural settings. One notable exception is the pinworm. In the United States, microbiologists estimate that 30% of all children and 16% of all adults are infected with pinworms. Although pinworms cause discomfort, they can easily be treated. Proper composting should destroy all of them.

Although one should not be carelessly cavalier about hygiene when composting feces, it's insane to go crazy about it. Do not fear all microorganisms; respect them, because the vast majority serve us well. Never lose track of the fact that feces usually only contain harmful microorganisms — ones that cause illness — when someone is carrying a disease or is visibly ill from one. Bear in mind as well, if someone's sick in your family with the flu or some other malady, you're far more likely to get sick from them sneezing on you or in your vicinity than from handling their waste — if you are careful, and who wouldn't be?

Although one should not be carelessly cavalier about hygiene when composting feces, it's insane to go crazy about it.

Be safe, but don't be paranoid. Handle buckets of urine and feces carefully. Cover them when you are transporting them to and from the compost pile to prevent contents from splashing out. Be careful when you wash the buckets and be careful with the rinse water. Cover up all the materials you dump in a compost pile. Remember: pathogens don't last long outside the human body.

Pathogens don't last long outside the human body.

Mother Nature's Clean-up Crew at Work

Bacteria and other microorganisms are vital for human existence. They're absolutely vital when it comes to recycling waste in ecosystems. They also assist us in a compost pile that "processes" humanure.

In a well-maintained compost pile containing humanure live a group of bacteria, known as *mesophilic* bacteria. They go to work immediately on the organic matter from composting toilets and sawdust toilets, devouring it with alacrity. Where do the mesophilic bacteria come from?

Well, frankly, from you.

Mesophilic bacteria originate in the human intestinal tract, notably our large intestines. Mesophilic bacteria break down organic matter, including organic cover material like sawdust and feces. During microbial decay, these organisms digest organic molecules, tearing them apart atom by atom. The energy stored in the chemical bonds that hold the atoms of organic molecules together is captured and converted to energy the bacteria can use. This process is not 100% efficient, however. In fact, it is far from 100% efficient. As a result, energy not converted into cellular energy used by bacteria is released as heat. This "waste" heat causes the temperature of the compost pile to increase rather rapidly to 111°F (44°C). Interestingly, this slight increase in temperature immediately begins to kill pathogens, if any, in humanure. They're adapted to surviving at body temperature of 98.6°F (37°C).

As temperatures rise, another group of bacteria takes over. These high-temperature bacteria, known as *thermophilic* bacteria, cause the temperature within the compost pile to skyrocket. In fact, the temperature inside a humanure compost pile can climb to 158°F (70°C). At such high temperatures, the remaining pathogens, if any, rapidly perish.

High temperatures persist inside a compost pile for a variable period — from a few days to a couple weeks. In some cases, they may last months. It all depends on ambient temperature and the amount of organic matter the bacteria have to dine on.

Temperatures need not rise so high to eliminate potential pathogens. In fact, a temperature of a mere 122°F (50°C) inside the pile for 24 hours will kill *all* pathogens. That includes pathogenic viruses, bacteria, fungi, protozoa, worms, and eggs of worms (Figure

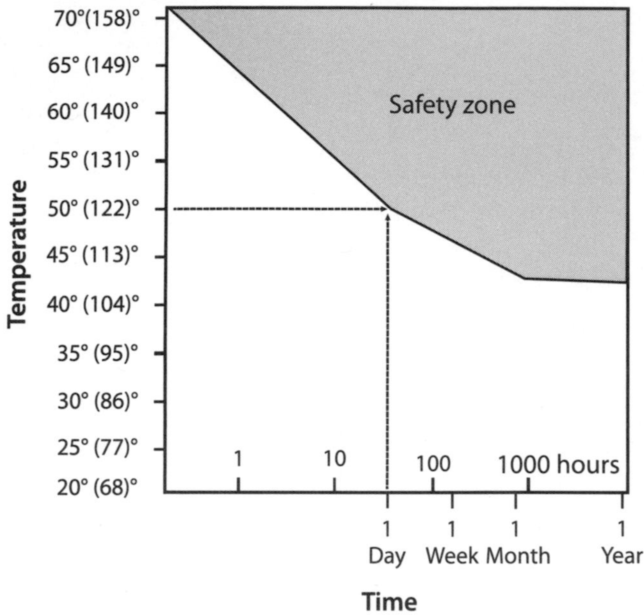

Fig. 5-5: Temperature, Time, and Bacterial Survival. *As you can see from this graph, the higher the temperature, the shorter the exposure time to kill potentially harmful bacteria.*

5-5). If the temperature isn't quite as high — say, only 115°F (46°C) — it may take a week to kill the pathogens, again, if any are present.

High internal temperatures are likely to occur very quickly on hot summer days, or even on mild days in the spring or fall. As a result, for most of the year, the compost pile in a temperate climate will naturally wipe out pathogens — if the pile is maintained properly. Proper maintenance means leaving the stuff in place and not disturbing the high-temperature zone. Keep adding material on top of it, and don't turn your pile — ever (contrary to instructions you may read in other books or online). You shouldn't need to. If you are adding sufficient fluffy cover material, the air spaces required for oxygen penetration will remain open. As you add new material, the thermophilic zone will soon expand into this material.

Once the thermophilic bacteria have had their fill, the temperature inside the compost pile begins to fall. At this point, their cooler-temperature cousins once again take over, decomposing what's left of the organic matter — especially the more resistant organic materials. Earthworms may also wiggle up into the pile, digesting what they can and leaving their own castings (a polite word for poo).

While most organic matter quickly succumbs to bacteria and other decomposer organisms, some of the organic materials such as wood shavings or sawdust are resistant to decay and therefore decompose more slowly. If you've studied botany, you know that woody materials contain cellulose, which decomposes fairly quickly. However, it also contains a complex organic compound called lignin that is considerably more resistant to decay. Lignin is attacked by a group of single-celled fungi, although they work a little slowly.

Although I've made the point several times in previous chapters, I'll say it again: While bacteria and other microbes quickly break down or kill potentially harmful microorganisms, it is best to "cure" humanure compost for a full year after the thermophilic stage. This additional time, as Joe Jenkins says, "adds a safety net for pathogen destruction." If any pathogens have survived the blast of heat, they will be destroyed in time inside the pile, gobbled up by some hungry critter, perhaps even an earthworm.

To improve the performance of your compost pile, do *not* turn it — that is, dig into and turn over the material. Turning a compost pile, say advocates of this practice, helps maintain adequate oxygen levels in the decomposing mass of cover material and human poo. Folks with lots of experience in composting, however, contend that it is not necessary to turn any backyard compost pile. I've never turned any of mine, and I've been composting since the mid 1970s.

Turning a humanure pile is not just unnecessary, it could be detrimental. That's because it could prematurely cool down the thermophilic zone, reducing the effectiveness of this natural pathogen-ridding process.

In the midst of a long, dry spell, feel free to sprinkle some water on top of your compost pile, but don't drench it. Decomposition occurs more rapidly when moisture levels are right. And always be sure you add coarse cover material such as straw or hay to help create air spaces inside the pile to maintain aerobic conditions. Your and your family's urine and excrement will provide moisture and nitrogen; straw and other organic cover material will provide the carbon that is required to ensure rapid decomposition. Don't be shy about adding

Mother Nature at Her Finest

Killing harmful bacteria, viruses, and other microorganisms is the job of your humanure compost pile, and it does a remarkable job of it — silently and effortlessly. It occurs right in your own backyard in suburbia or out in the country. It's Mother Nature at her finest.

leftovers from the kitchen or yard trimmings, leaves, shredded cardboard, shredded paper, or cotton clothing.

How long it takes for a humanure compost pile to top off (fill a compost bin) depends on many factors. For Jenkins and his family, it takes a year. When the first bin in his three-bin system is full, he starts a second one, leaving the first to "cure" (decompose) while the second pile receives its deposits (Figure 5-6). At the end of year 2, the nutrient-supercharged organic compost in the first bin is ready to use. At this point, it should be crumbly and easy to dig into soil.

If you live in a cold climate, don't worry: you can deposit organic materials from a composting or sawdust toilet throughout the

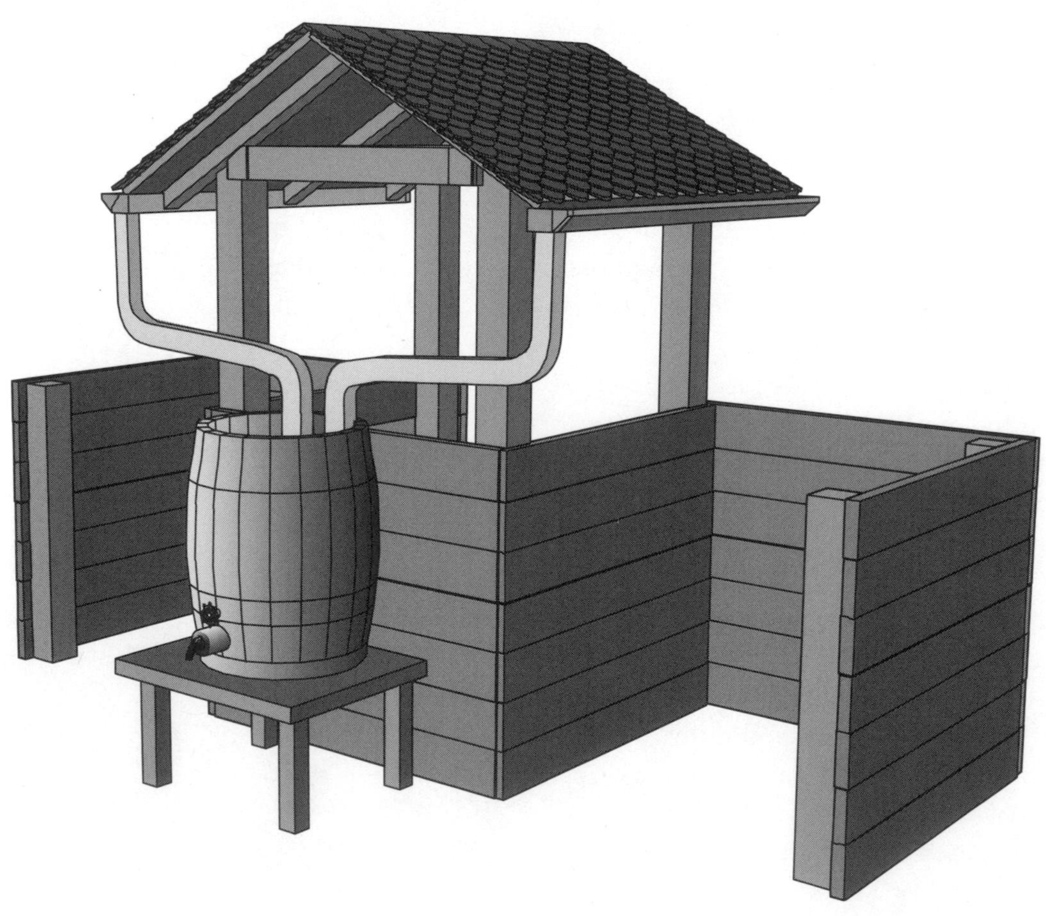

Fig. 5-6: Jenkins' Three-Bin Compost Bin. *Note that the middle bin is covered to protect cover material.*

winter. As in the case of an outdoor composting toilet, the contents may freeze and remain frozen all winter in really cold regions. In milder regions the compost pile may freeze and thaw several times. Either way, there's no harm done. Just keep adding material. Rest assured, when warm weather returns, your microbial co-conspirators in the cycle of life will resume work. The pile will heat up and decomposition will re-commence.

Thriving under Benign Neglect

I'm sure there are compost experts who will disagree with my *laissez-faire* attitude toward composting, but I've never turned a compost pile since I started organic gardening in the late 1970s. What is more, I have enjoyed great success over the years — even in arid Colorado.

I simply add organic matter from various sources, sprinkle in a little topsoil, and let Mother Nature to do her magic. In the really dry months in Colorado, I sprinkled my compost piles with a little water, when I remembered, and all of my compost piles have thrived under my benign neglect.

So, resist the urge to mess with your compost pile. Just let the stuff sit and decay. Allow the bacteria and other microbes to do their magic. They've been doing it for eons and will get the job done without complaint — or a paycheck.

What Should You Do with the Composted Poo?

On our farm, we routinely add well-aged composted cow manure to our gardens and greenhouse, where it helps us raise an abundance of vegetables.

We treat well-aged and thoroughly composted humanure identically. I can't formally recommend this strategy. It's illegal in many places to add composted humanure to vegetable garden soil, and the literature on composting humanure is filled with cautionary advice. Most folks, in fact, recommend applying composted human waste to flower gardens or orchards. They advise readers to dig it deep into the soil to prevent any problems.

Here's the deal, though: if your humanure compost pile has reached the thermophilic stage or remained at high temperature for an extended period, as noted earlier, and you've let the waste decompose for a year after it has come out of your compost or sawdust toilet, it should be perfectly safe. You should be able to work it into all your gardens — not just your flower gardens — with a hoe, rake, or a cobra — one of the greatest, if not the greatest, garden tool ever invented (Figure 5-7). That's what Joe Jenkins does.

Fig. 5-7: Friendly Cobra. *This is an amazing garden tool, ideal for digging in soil and working humus into your soil.*

And I'm sure there are hundreds of other rebel humanurists who do the same.

Check with local regulations, if any. If there are regulations, they will probably require you to bury humus — deep in gardens that are not used for raising edible produce.

One way to find out what regulations are is to call your local state or county health department. Tell them that you are considering a composting toilet and were wondering what regulations have been adopted for the safe "disposal" of composting toilet humus.

I'd not mention your plans to set up a humanure compost pile in your backyard, however. It's not a bad idea to leave neighbors out of the loop, too. Government officials and neighbors can get pretty weird pretty quickly about earth-friendly behaviors that don't fit within the narrow confines of normality. Skittish neighbors might even turn you in to authorities. See the accompanying textbox for an account of my personal experience with skittish neighbors.

Conclusion

To paraphrase Joe Jenkins — the man who popularized the simple and effective sawdust toilet — people who compost their waste recognize that recycling humanure is one of the regular and necessary responsibilities for sustainable human life on this planet. Recycling human excrement is a noble cause done out of respect and a deep understanding of how we have to act to protect and sustain our place in the cosmos. There are other ways to contribute positively, such as greywater recycling. I urge you to read on to consider other positive steps you can take to build a sustainable future.

Where's the Smell Coming From?

When I was an assistant professor of biology in the 1970s at the University of Colorado in Denver, I maintained a huge compost pile in my garden. I used it to reclaim nutrients from weeds from the garden, leaves, and kitchen scraps. No humanure.

One summer day I came home from teaching summer school to find a city official snooping around my backyard. When I asked him what he was doing, he told me that a neighbor had reported an odd smell. They thought it was emanating from my compost pile.

When I asked what he'd found, he said nothing. He told me that the smell my neighbor had detected was most likely from the city's own sewage treatment plant located about a mile north and west of our neighborhood. The wind must have drifted our way, alarming my neighbor and sending her scurrying to the phone to turn me in.

My advice is to follow the rules, but fly under the radar. You don't want people snooping around your backyard looking for excuses to shut you down. Do your thing and don't try to convert everyone in the neighborhood.

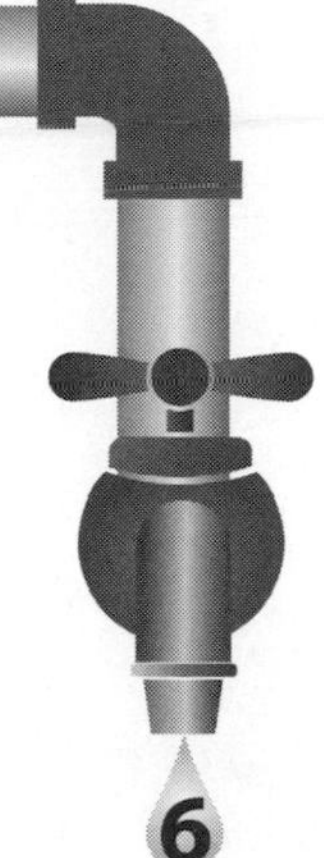

Capturing and Capitalizing on Greywater

Every time you brush your teeth, wash dishes, or shower, you are adding to a massive out pouring of "waste(d) water" that streams out of our homes throughout the world each day. The product of these seemingly innocuous actions is billions of gallons of slightly soiled water, called *greywater*. Its name derives from the color it acquires from soap, detergent, shampoo, cream rinse, skin oil, and dirt.

Greywater is the effluent from bathroom and utility sinks, showers, bathtubs, automatic dishwashers, and washing machines (except when diapers are laundered). "Waste" water derived from flush toilets is referred to as blackwater, although, brown water would be a more accurate description. Water from kitchen sinks is also referred to as blackwater by some jurisdictions, rather than greywater. That's because it may contain potentially harmful bacteria from meats washed during meal prep. If you use a garbage disposal, kitchen sink effluent could also contain meat scraps and other organic debris. To take into account its slightly different composition, greywater aficionados often refer to the effluent from kitchen sinks and washing machines when laundering dirty diapers as "dark greywater."

Greywater can be used as is to irrigate indoor and outdoor plants. It can also be purified and used to flush toilets. To do so, however, requires a very high level of purification. The water must nearly be drinkable. If purified even more, it can be made suitable for bathing and human consumption. That's what goes on in

outer space in the International Space Station. That said, the vast majority of greywater systems in use today are designed to irrigate.

Greywater comprises 50% to 80% of the water we send down our drains every day, according to greywater expert Art Ludwig of Oasis Design. North Americans produce billions of gallons of greywater every year.

In conventionally plumbed homes, greywater and blackwater are combined (Figure 6-1). That is, sink and shower water join toilet water. In rural homes, the combined grey and blackwater generally flows into a backyard septic tank. In cities and towns, grey and blackwater are piped to municipal sewage treatment plants. (Rural residents who live close enough to cities and large town are often served by sewage treatment plants.)

As noted in previous chapters, water from composting toilets (blackwater) contains a variety of valuable plant nutrients, such as

Fig. 6-1: Common Household Plumbing. *This illustration shows how drains and vent pipes from common household fixtures — sinks, showers, and toilets — are installed. Note that in a typical home, blackwater and greywater are combined.*

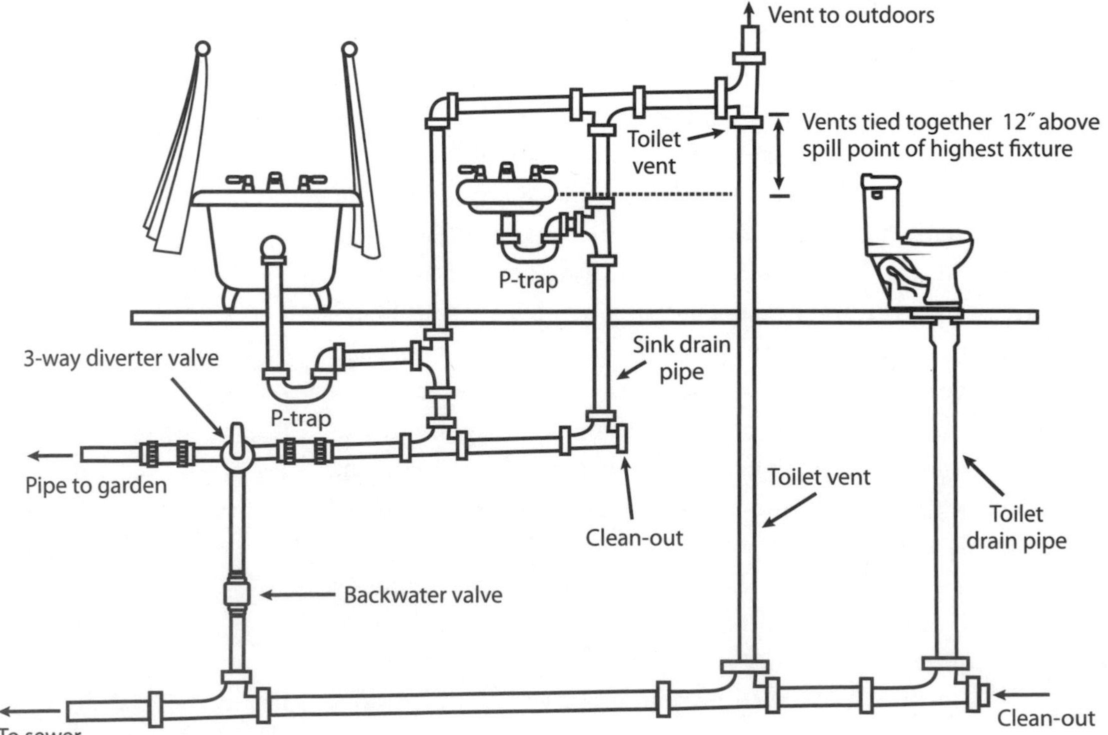

A Word on Art Ludwig

Throughout this chapter, you'll hear a lot about Art Ludwig (Figure 6-2). Inarguably, Art's the pre-eminent international expert on greywater. His books, DVDs, and website contain a wealth of valuable information on designing and operating greywater systems. Art's available for consultation as well. Before you design and build a system, I highly recommend reading the most recent edition of his book, *Create an Oasis with Greywater*, and viewing his DVD "Laundry to Landscape." You can purchase them through his website, Oasisdesign.net. By purchasing directly, you'll save money and help support Art's extremely valuable work.

Fig. 6-2: Art Ludwig. *Art's life mission has been to perfect and promote integrated, optimized systems for water, wastewater, energy, food production, etc.*
Courtesy of Art Ludwig

Fig. 6-3: Indoor Planter. *The planter in my earth-sheltered rammed earth tire and straw bale home in Evergreen, Colorado, has been irrigated by greywater from my washing machine for the past 15 years.*

nitrogen and phosphorus. It also contains organic matter that enriches garden and agricultural soils, assisting us in growing our own food more sustainably. In contrast, greywater is rather nutrient poor. Even so, its nutrients can be captured and put to use. It is also a great source of water that can be used to irrigate all sorts of plants. I have used greywater in my home in Colorado to water plants in my indoor planters as well as my vegetable garden since 1996 (Figure 6-3). I have even used it to water trees that I planted around my home.

Ironically, most people irrigate their lawns and gardens with highly filtered, chemically sterilized drinking water, while flushing thousands of gallons of perfectly usable greywater down the tubes each month. You can help put an end to this waste.

In this chapter, we'll explore ways to capture massive amounts of wasted water and put it to good use irrigating vegetation around our homes and offices. As in other chapters, I'll help you understand how these systems work, what some of your options are, and some things to look out for. Be sure to study this topic in more detail if you decide to design and install a greywater system. There's a lot to know and things change rapidly. Ludwig's materials should definitely be consulted, as he keeps up on all the recent information and revises his books often to incorporate new material.

The Human Hydrologic Cycle

Before we delve into greywater systems, let's take a look at where our water comes from and where it goes after we're done with it. This will help you understand the importance of greywater capture and reuse and the beneficial impact you will have by recycling greywater.

If you live in a city or town, that water you use in your daily life — for drinking, flushing toilets, washing clothes, showering, bathing, and watering lawns and gardens — is typically drawn from rivers, lakes, or groundwater "reservoirs," known as aquifers. In coastal regions, water is sometimes drawn from oceans, desalinated, disinfected, and then distributed to residents and businesses.

Water from all sources is pumped from its source to a water treatment plant where it is filtered and treated with chlorine or similar chemical compounds to kill potentially harmful waterborne microbes, mostly bacteria. Sterilization helps protect us and helps keep the pipes in the water distribution system from biofouling — that is, becoming lined with bacterial colonies. Fluoride is added in some jurisdictions to harden enamel and prevent tooth decay.

This highly purified, well-sterilized drinking water is then delivered to our homes and places of work, worship, play, and entertainment via an elaborate set of underground pipes. The vast majority of the purified drinking water delivered to end users, however, does not require the high levels of purification we employ. Watering grass, for example, doesn't require chlorinated water. Nor does flushing toilets. Rather than building two water delivery systems, however, cities and towns have opted for one system that provides relatively pure water (potable) for all uses.

In rural areas, water is typically drawn from underground deposits, known as aquifers. It is usually pumped out of the wells via submersible pumps. The pumps are suspended inside a pipe deep inside the well casing (the pipe that lines the well). Well water is pumped from the well to a pressure tank inside our homes. This device keeps the water under pressure at all times, so when you turn on a faucet, water comes out. From the pressure tank, water is distributed through pipes to sinks, showers, washing machines, dishwashers, toilets, and other fixtures.

What Are We Doing?

Ironically, most people water their lawns and gardens with highly filtered, chemically sterilized drinking water, while flushing thousands of gallons of perfectly usable greywater down the drain each month.

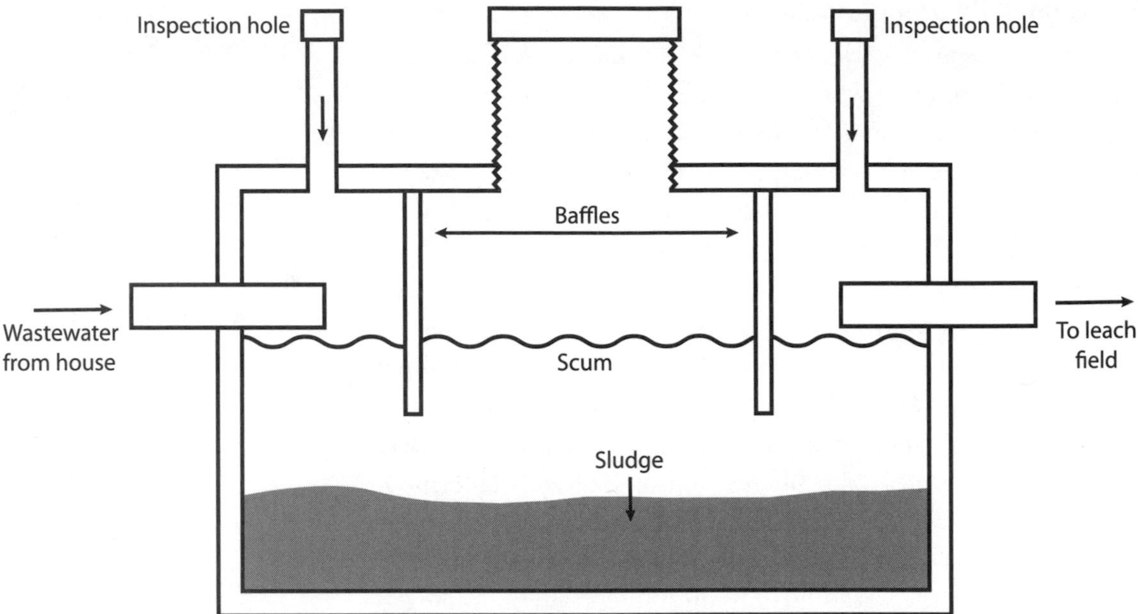

Fig. 6-4: Septic Tank. *Septic tanks are typically plastic or concrete tanks that allow solids to settle to the bottom. Liquids are siphoned off the top of the mix and subsequently flow into leach fields.*

Inside our homes, water tends to be quickly used, then immediately disposed of. In cities and towns, the water is piped to sewage treatment plants. In rural areas, it is typically piped out of our homes to septic tanks buried near our homes.

In a septic tank, solids such as feces and hunks of food washed from dishes or flushed out of the garbage disposal precipitate out — that is, they settle to the bottom of the septic tank, forming a dark brown sludge (Figure 6-4). Sludge builds up over time, requiring periodic removal. Sludge is pumped out and usually hauled to municipal sewage treatment plants.

The liquefied waste in septic tanks drains out of the side of the tank into a set of perforated pipes buried in the ground, collectively known as a leach field (Figure 1-5). The perforated pipes of the leach field are buried in trenches filled with crushed rock, then covered with dirt. Leach fields are designed to allow effluent from the tank to seep into the ground. However, because the pipes are buried fairly deep within the ground, very little, if any, of the dissolved organic waste and inorganic nutrients released into a leach field re-enters the nutrient cycles and is put to good use.

Most of it seeps into soils and surface groundwater. If cracks in the strata beneath the superficial groundwater aquifers allow waters to drain into deeper aquifers, septic tank effluents can pollute them, contaminating our drinking water. (See textbox for more details.) However, as you will learn in Chapter 7, effluent from leach fields can be routed to plant roots through a branched drain network and subsoil infiltrators. This system, dubbed a "green septic," by inventor Art Ludwig, helps prevent groundwater pollution and puts a valuable resource to good use.

Now that you understand how water is delivered and "waste" water is disposed of, let's take a look at greywater systems.

Rules of Safe Greywater Use

Before we explore some of the most popular greywater systems in use today, let's review some rules of safe, hygienic greywater capture and reuse that I have gleaned from Art Ludwig's website and books, my years of experience with greywater, and a number of other sources.

The first rule of any greywater system is to avoid ingestion or human contact. Design your system to minimize or eliminate the possibility of human contact — or contact with pets. To do so, deposit greywater below the surface of mulch around vegetation or below the soil surface, in the upper portion of plants' root zones. In some areas, such as Arizona and New Mexico, greywater can be released from pipes directly over a highly porous substrate, like thick beds of mulch or wood chips surrounding trees or shrubs. Figure 6-5 shows some options that prevent greywater from pooling on the surface.

Be sure to prevent greywater from pooling on the surface. The main concern with greywater pooling is that children might play in the water or cats and dogs might drink it. Many kids — especially boys — find puddles entertaining, and could end up exposing themselves to potentially harmful microorganisms. Even though greywater is relatively clean compared to blackwater, water exiting our showers, baths, and washing machines may contain small amounts of feces washed off our bodies or undergarments.

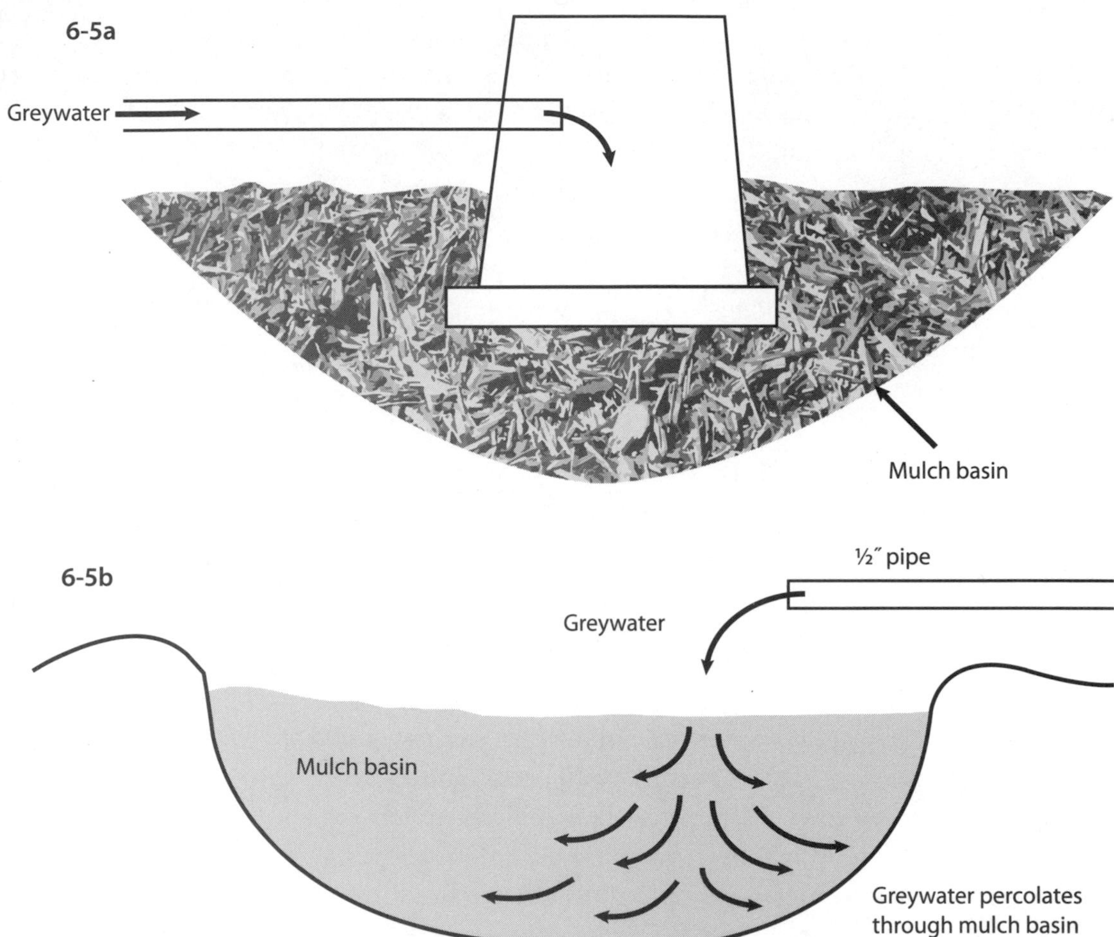

Fig. 6-5: Mulch Basin. *Greywater can be deposited into a mulch basin in a couple of ways. (a) The mulch shield protects the outlet and ensures rapid percolation into the ground. (b) In some states, greywater can be deposited directly onto mulch without a mulch shield.*

Feces may contain potentially harmful microorganisms. That said, it should be pointed out that the risk of someone becoming ill from greywater is extremely low. In fact, according to Art Ludwig, "in one billion system-user years of exposure over the past six decades of accurate disease tracking in the United States, there's never been a documented case of illness contracted by exposure to greywater!" Ludwig adds, "It's easy to understand why. When you are breathing the same air, using the same kitchen and bathroom, kissing one another, changing diapers of your kids ... the added exposure from bathwater you just shared going onto a tree in the year is minimal." "That said," says Ludwig it's still not wise "to build pathways for disease transmission into your system."

It should be obvious that one should avoid drinking greywater or allowing it to contaminate drinking water. Also avoid direct contact with foods that are eaten raw, such as spinach, chard, lettuce, bok choi, or carrots. "Beyond that," says Ludwig, "there is not a lot of risk from greywater systems."

The second rule of safe greywater use is not to store untreated greywater for more than 24 hours. Although greywater contains very little organic matter, it turns putrid if left to sit for more than 24 hours. That's because the naturally occurring bacteria in greywater rapidly consume the organic matter and, in the process, release some rather foul-smelling chemicals.

The best (meaning low-odor) way of handling greywater is to use it immediately — for example, to pipe greywater from various sources such as washing machines and sinks directly out onto outdoor plants. The other option is to drain the effluent into a nearby holding tank, and then either immediately pump it out to or let it flow by gravity to outdoor plants so the tank empties quickly.

Preventing odor from developing in holding tanks can be a big challenge. Be sure all the water is completely drained from the tank during each cycle. If you place the drain in a barrel a few inches above the bottom of the tank, greywater will never fully drain. Water remaining in the bottom of the tank will sour on you overnight and begin emitting a foul odor. I have found that a fairly tight lid contains odors, but it is best to avoid this problem by ensuring that the tank drains fully after each cycle. Or, you can pump the water immediately to a greywater distribution system that irrigates trees and other plants. More on this shortly.

Third, be sure that greywater can't flow into and hence contaminate nearby ponds, lakes, and streams — even small perennial streams or drainage ditches.

To prevent greywater from flowing into nearby water drainage, most greywater aficionados create mulch basins around plants they are irrigating, as shown in Figure 6-5. Mulch basins must be filled with a material that creates a porous substrate, such as tree bark, wood chips, and straw — or some combination of them. Bark and similar organic mulches allow the effluent to quickly drain into underlying soil and to the root zones of plants.

A fourth requirement of virtually all greywater systems is the installation of a diversion valve such as the one shown in Figure 6-6. Diversion valves are used to divert water from the greywater irrigation systems to one's septic tank or to the sewer. Diversion may be necessary during periods of heavy rain or when the washing machine is used to launder feces-laden diapers. It could also occur when your house suddenly becomes inundated with guests and there's a lot of shower water going down the drain, and you risk flooding your plants, although this probably won't be a problem if you have designed your system correctly.

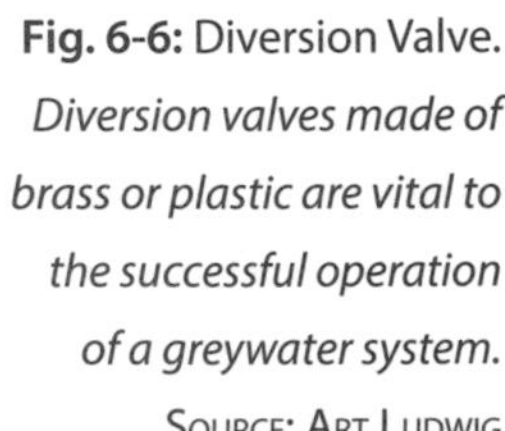

Fig. 6-6: Diversion Valve. *Diversion valves made of brass or plastic are vital to the successful operation of a greywater system.* Source: Art Ludwig

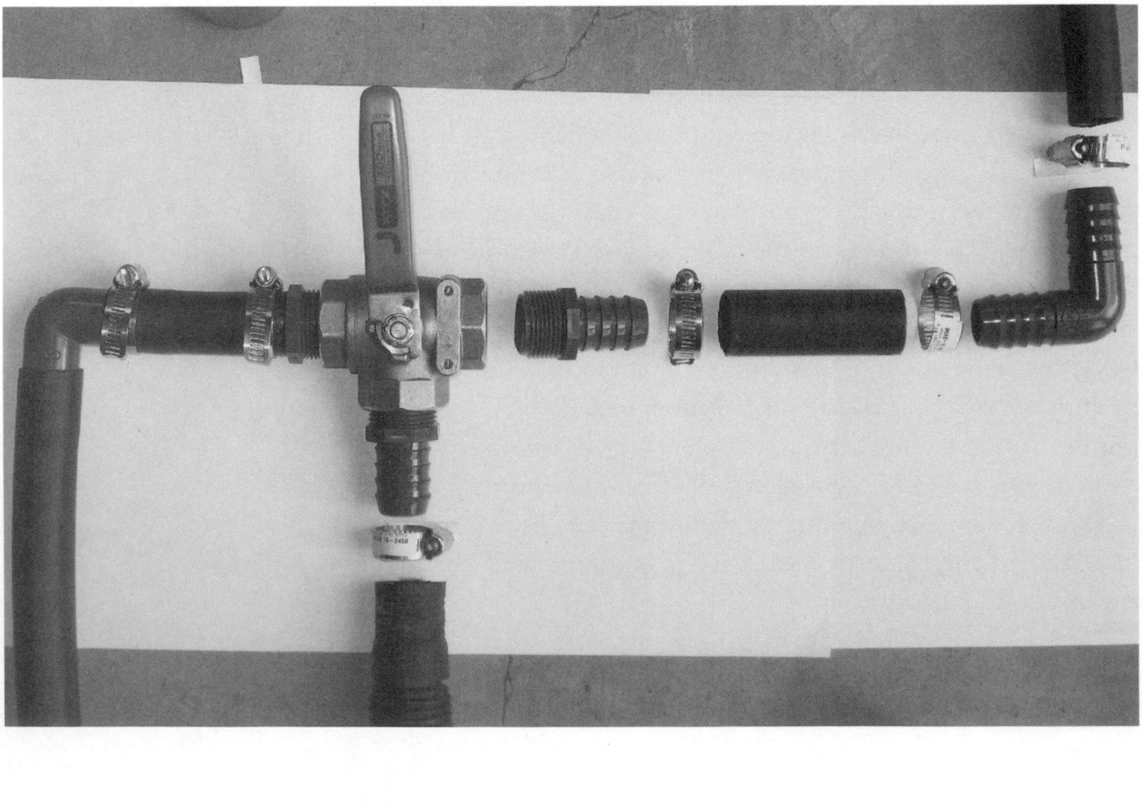

The diversion valves used by most people are three-way pool valves, described in more detail shortly. They're available through

pool supply outlets and online (of course!), but in my experience generally not through plumbing supply houses or home improvement centers.

A fifth rule of successful greywater system is to design a system that minimizes or eliminates clogging, thus reducing or eliminating periodic cleansing. Greywater from washing machines contains tiny fibers from clothing (that's one reason your clothes wear out). It may also contain hair. Hair and cloth fibers clog filters (if any) and small openings in greywater distribution systems, for example, drip irrigation emitters (devices that release greywater at each plant). Organic matter and bacteria that grow in greywater can also clog openings. They form gooey deposits that stop up small pipes. Tiny openings in emitters can quickly become blocked with this organic crud. I discovered this when establishing my first greywater system in the late 1990s. I hooked up a greywater tank supplied by my washing machine to a drip irrigation system I designed to water my garden. Within a week of hooking up the greywater, very little water was dripping out of the porous rubber tubes of my irrigation system. They had become clogged with fibers and organic gunk.

To prevent greywater goop from building up in the mulch near the end of each pipe, terminate the pipe in a hollow chamber within the mulch zone, several inches (about 7 cm) above the top of the mulch, as shown in Figure 6-5. Greywater flows freely out of the end of pipe or emitter (valve). It then drops a few inches onto the mulch in the basin and rapidly seeps out of sight below the mulch and then into the surrounding soil where bacteria quickly decompose organic matter.

As shown in Figure 6-5, an outlet shield for mulch basins can be made by drilling a hole in the side of a plastic pot large enough for the pipe to enter. More on these shortly.

The sixth rule of successful greywater use is to match greywater delivery to the water requirements of your plants. Some plants like a thorough wetting, followed by a dry period. Others like being immersed in water for long periods. Still others prefer drier conditions throughout the year. In addition, water requirements may vary during the growing season. Further complicating things, the amount of rain falling on one's yard often varies during the year.

Matching greywater delivery to variable water requirements is one of the most difficult parts of designing and building a greywater system. It requires a knowledge of plants that few of us have and careful monitoring of rainfall. For advice, it's best to contact a horticulturalist or go online and read up on the trees and plants you will be irrigating. Be sure to read Art Ludwig's book, *Create an Oasis with Greywater* as well. Art has included valuable worksheets and advice that can help you figure all this out. Advice on protecting roots from being constantly immersed in greywater will be presented later.

The seventh and final rule of successful greywater system design is to keep them as simple as possible. Let gravity do as much work as possible delivering water to your plants.

The simplest and best way to get greywater to the yard is from a washing machine using the "Laundry to Landscape" system developed by Art Ludwig, discussed shortly. This system utilizes the washing machine's own pump. It removes water from the washing machine after each wash and rinse cycle. Normally, it pumps the wash water it into a nearby utility sink or into a drain located in the wall behind the washer. To capture greywater, the washing machine hose can be attached to the greywater distribution pipe.

For greatest success with greywater, then, the rule is this: the simpler, the better. Even so, there are several ways to screw up a simple system, and tiny errors can mean

Rules of Successful Greywater Systems

- Don't store untreated greywater for more than 24 hours.
- Prevent contact with greywater.
- Prevent greywater from flowing into surface waters.
- Install a diversion valve so greywater can be diverted into the septic/sewer.
- Design the system to prevent clogging.
- Match greywater delivery to your plant's requirements.
- Keep greywater systems simple.

Source: Joe Jenkins, *The Humanure Handbook*

the difference between success and failure. So, study the material in this chapter very carefully, then delve into the topic in more detail. Be careful where your information comes from. The Internet is not always the best source of information — unless you seek the most reliable sources, like Art Ludwig. Frankly, I'd start with his online and published material.

The list in the accompanying textbox summarizes the principles of successful greywater design and construction. You may want to take a minute to review it.

With these basics in mind, let's look at some system designs that you can start working on today.

Designing and Building a Greywater System

Greywater systems range from basic to complex, but most are fairly simple and easy to understand. If you have some experience in plumbing and some knowledge of home construction, they're not generally difficult to install either. You will also find that the plumbing skills required to work with plastic pipe and pipe fittings are fairly easy to master. It's far from rocket science.

The biggest challenge you'll have is figuring out which parts you'll need and what they are called — and then finding them. Fortunately, there are some excellent sources for designs. As you might suspect, the very best and most current source of information is greywater guru Art Ludwig's books and his DVD available through his website (www.oasisdesign.net). Art updates his material with great regularity, so it's wise to seek information through his video, books, and website.

Another valuable source is the City of San Francisco's *Greywater Design Manual for Outdoor Irrigation.* It draws heavily from Art's extensive work; the city's rebate program is entirely based on systems Art developed.

You can also find lots of information on greywater systems, including photos, at the website of Greywater Action (greywateraction.org). Study all this information, then design and install your system. After careful study, you should be able to build your own system.

It's always a good idea to start small and to enlist the aid of folks who have more experience. Proper design means sizing your system properly to provide the right amount of water to each plant. I'll provide a few general guidelines, but once again, go to the authorities on the subject.

In this section, we will explore a couple major greywater system options and describe pertinent details of their design and construction. You may be tempted to purchase a commercially available greywater system after reading this material, but remember these systems can be pretty elaborate, expensive, and generally don't last long or provide satisfactory performance in the field, according to Ludwig. They generally come with a tank, filter, pumps, and valves. You will still need to modify the plumbing in your home to "feed" the system. What is more, filters need to be regularly cleaned and get pretty smelly, pretty quickly. As a result, it's not a popular job among family members or employees. More important, commercial systems are designed to purify greywater — that is, remove most, if not all, of the contaminants. That means they remove the nutrients we can deliver to our plants. Because I suspect most readers are pursuing greywater to irrigate vegetation and won't need to purify greywater, I won't discuss commercial systems in this book in any depth.

If all this seems like too much, you may be able to hire a professional to design and install your system — especially if you live in a state where greywater systems are popular, such as California, New Mexico, Arizona, and Texas. Be sure to hire someone who's knowledgeable and enthusiastic about helping you, not a skeptic or a detractor — a plumber who will do it for you, though he or she thinks you are nuts. A word of caution, however, when it comes to engineers, plumbers, and permitted systems: professionals tend to favor complex, highly mechanized, over-designed, energy-intensive, and costly greywater systems, according to Ludwig. Simplicity, as noted earlier, often works best.

With this in mind, let's take a look at the most basic system design, known as a laundry drum system. Although the system is antiquated and rarely used, I present it to help you understand the evolution of greywater systems and how these systems work.

Designing and Building a Laundry Drum System

The easiest and most economical system to install is the laundry drum system, shown in Figure 6-7. This system was developed by none other than Art Ludwig.

These systems operate by gravity and hence are ideal for homes in which the washing machines are located in a laundry room on the first (or possibly second) floor of the house, well above the irrigation field. In warmer climates, the washing machine could be located outside on a covered patio or an enclosed back porch. The drain hose from the washing machine empties into a 55-gallon plastic barrel located nearby — typically outside the house. The washing machine hose can be run through an open window or directly through an exterior wall.

In this system, the 55-gallon barrel serves as a storage tank, henceforth referred to as a *surge tank.* Its name derives from the fact that the tank absorbs the sudden outpouring of water produced as a washing machine empties after each wash and rinse cycle. Attached to the bottom of the tank is a spigot to which a garden hose is attached. The surge tank can be placed on cement blocks or a wooden platform to provide easy access to the valve.

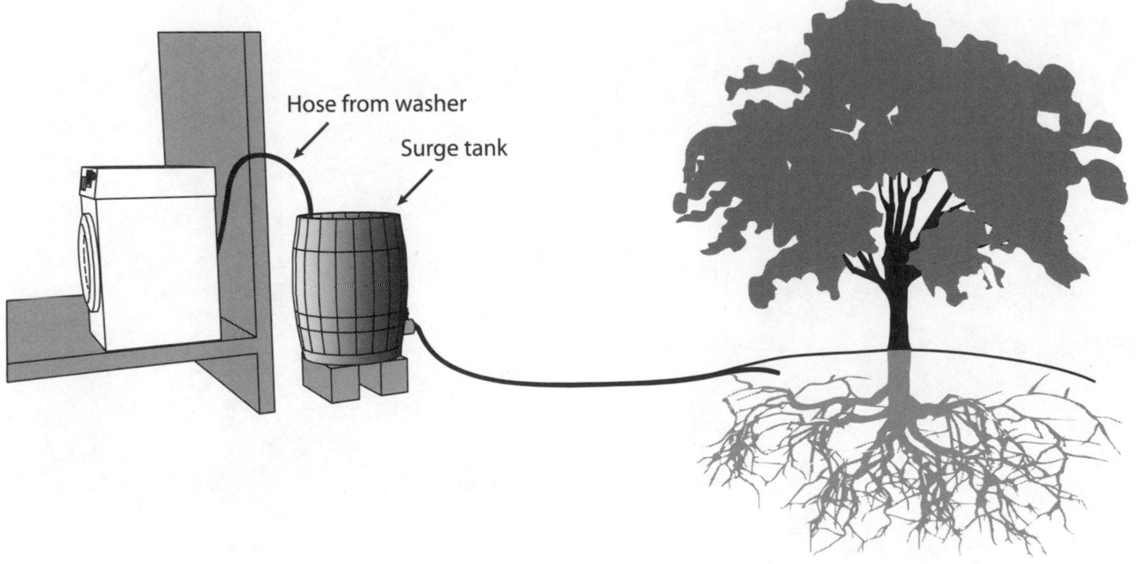

Fig. 6-7: Laundry Drum System. *In this simple system, greywater is pumped from the washing machine to a surge tank, then flows by gravity to trees or bushes.*

The garden hose can be run to plants near the house and moved manually from plant to plant (actually mulch bed to mulch bed around each plant) after the washing machine has emptied. To save time, you can install a hose splitter — or several of them. Splitters will allow you to divert water to several areas at the same time, saving time and energy (Figure 6-8). If you are manually

Fig. 6-8: Hose Splitter. *Hose splitters like those shown in Figures a, b, and c allow you to water two or more areas simultaneously and eliminate the need to drag a hose from one location to another. I recommend metal splitters over plastic ones for durability.*

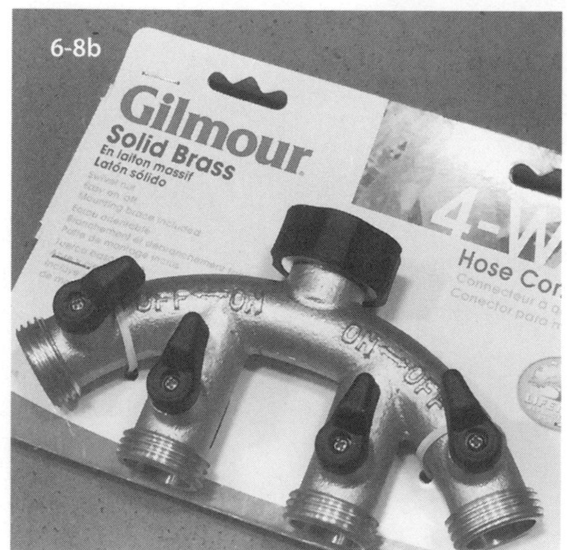

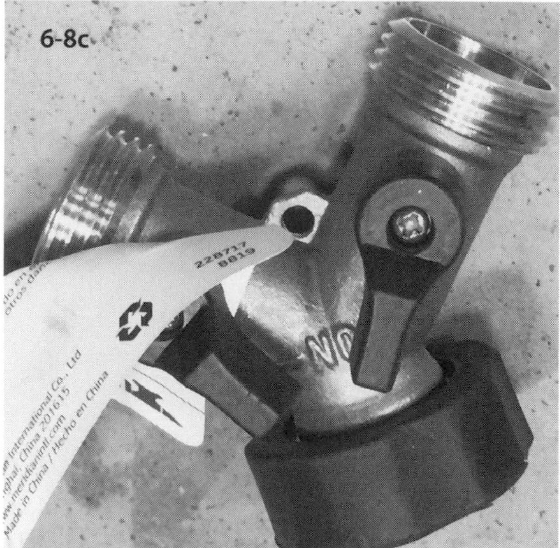

moving the hose, be careful not to splash greywater on yourself or on the ground.

In the laundry drum system, water flows by gravity down the hose — or hoses — to plants at a rate that should prevent flooding and pooling. Be sure hoses run downhill.

To divert wash water from plants during periods of heavy rain or during cold winter weather, simply detach the hose from the washer and place it in the standpipe — that is, the drainpipe into which it would otherwise empty. The standpipe is typically located in the wall directly behind the washing machine. If your clothes washer empties into a utility sink, move the hose there. Either way, water will now flow to the septic tank or sewer.

Also be sure to divert water away from plants when you are using bleach. Chlorine in bleach is harmful to soil microbes as well as plants. If you are washing soiled diapers, divert the wash water to the sewer or septic tank to avoid fecal contamination.

Pros and Cons of Laundry Drum Systems

Laundry drum systems are as simple as they get. They require very few materials, are inexpensive, and are relatively easy to install. For very little time, effort, or money, a homeowner can create a crude, yet effective greywater irrigation system. It'll lower your water consumption, reduce the cost of water, help protect waterways and groundwater, and save energy.

The biggest downside to this system is that you'll need access to a nearby window in the utility room — and you'll need to keep it open a few inches when the washing machine is in use. If insects are a problem, stuff a towel in the opening to prevent bugs from joining you, or having you for dinner. You can detach the hose after each load of laundry and shut the window or the window screen to minimize unwanted guests.

Laundry-to-landscape System

Figure 6-9 shows a more sophisticated — and very popular — greywater system referred to as a laundry-to-landscape system. It supersedes the laundry drum system, and its installation may

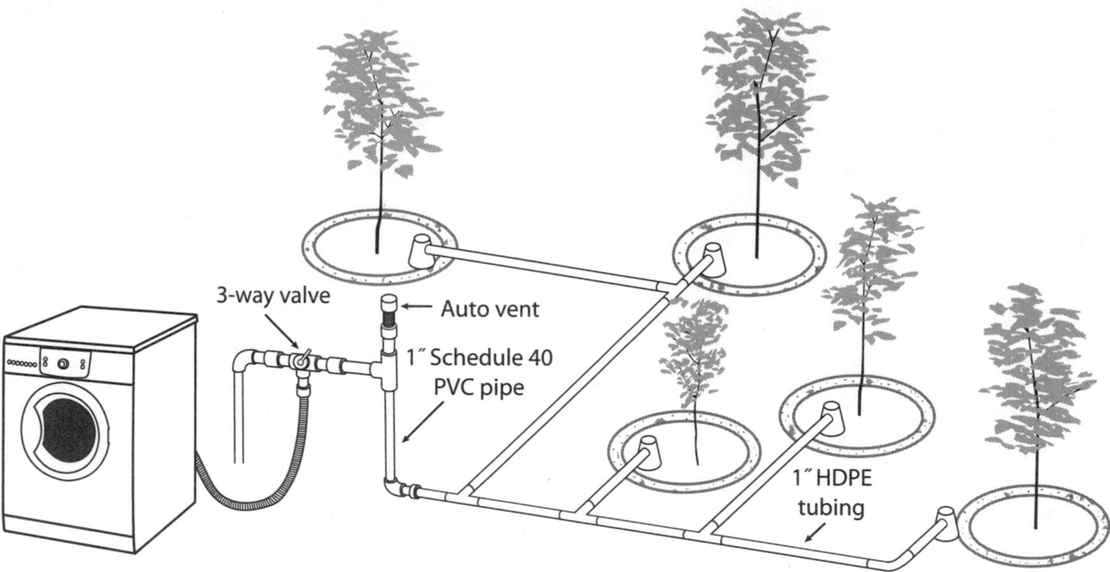

Fig. 6-9: Laundry-to-landscape System. *This slightly more sophisticated system highly automates the process. These systems must be equipped with anti-siphon devices.*

qualify you for rebates from city water departments. As illustrated, it is also designed to deliver water from the washing machine to vegetation in one's yard.

This system was invented by greywater guru Art Ludwig, and published without patent for release into the public domain. It has become an extremely popular system. What is more, dozens of cities now offer rebates for the installation of Art's system. (Many thanks to Art and others like him who work tirelessly to help create a more sustainable way of life!)

Laundry-to-landscape differs in several key respects from the antiquated laundry drum system. One of the key differences is that there's no surge tank. The washing machine hose empties under pressure from the washer pump into a pipe that directly runs to the drain field. Water flow can also be diverted to the septic system or sewer. Which direction greywater flows is determined by adjusting a three-way valve, shown in Figure 6–9.

Another key difference is that greywater runs from the washing machine to the irrigation field directly through the wall. Because of this, you don't have to leave the utility room window open when washing clothes and watering plants. You can see why this is a popular option for many homeowners!

Water is typically piped out of the house in one-inch Schedule 40 PVC pipe. Schedule 40 is a heavy-duty pipe used in plumbing. ("Schedule 40" refers to the thickness of the wall of the pipe. The one-inch designation refers to the inside diameter [ID] of the pipe.) Do not use the less-expensive, lighter-weight, and much more fragile sewer and drain pipe. (It's referred to as Schedule 35.) It is fairly brittle and won't last nearly as long. If the drain pipe is exposed to sunlight outside the home, be sure to paint it with latex paints or instead buy grey plastic (PVC) conduit used for electrical work. Latex paint prevents photodegradation and makes the system look prettier, especially if the paint matches or compliments the house's color. Grey PVC is also UV resistant.

A Word on Valves

The three-way valve is vital to the success of a laundry-to-landscape greywater system. Select a high-quality valve. Your best bet is a three-way brass valve. Be sure to mount the valve above the flood rim of the washing machine (the top of the drum). The valve should also be positioned slightly above the drainpipe that carries water to the sewer or septic tank. That way, water will flow down and out of the house when diverting water to either of these endpoints. Also be sure to install the valve so you can easily reach and operate the handle. As Art Ludwig noted in a private communication, "If you never use toxic cleaners and your yard is never saturated, you can skip the valve and just run laundry water to the landscape full time."

The one-inch PVC drain pipe typically empties into flexible one-inch high-density polyethylene pipe, typically called HDPE. HDPE can be run underground or along the surface to the mulch basins around your plants. To reduce the flow of water to smaller plants or plants with lower water demands, step the one-inch HDPE pipe down to ½-inch HDPE. Other options to reduce flow are also available, and will be discussed shortly.

When running pipe along the surface, be sure it is secured and protected so people won't trip over or damage it. If a pipe crosses a path in the garden, it's a good idea to bury it under a few inches of dirt or mulch. Use heavy-duty landscape staples to secure pipe to the ground. If the pipes cross grassy areas, it's best to run them underground to prevent tripping and to avoid damage from lawn mowers and other equipment.

If pipe is buried throughout the system, be sure to maintain a continual decline of ¼ inch per foot for optimal flow. If you don't, water could pool inside the pipes and not distribute evenly. If you want to water one area at a time, be sure to install diversion valves in the pipe at key locations to control flow.

In laundry-to-landscape systems, the electric pump inside the washing machine propels water through the tubing along sections of pipe that run across flat areas of your yard. Be sure not to "ask" the washing machine pump to move the water very far, more than 100 feet, say, or more than a few feet uphill. If you do, you'll very likely damage the pump by creating too much backpressure on it.

Laundry-to-landscape systems can be operated year round in warm climates, but will very likely need to be temporarily shut down in areas where freezing occurs. This terminates the flow of water to plants when they are dormant and prevents pipes from freeze damage. If freezing temperatures are rare and you want to continue to supply water to your plants all winter, check out Art Ludwig's DVD. In it, he explains various ways to design systems to prevent greywater in pipes from freezing.

Drain pipe in a greywater system creates a siphon. (The drain field lies below the tank of the washing machine.) Because of this, greywater systems can siphon water out of a washing machine as it fills. To prevent this annoying problem, be sure to install an anti-siphon valve like the one shown in Figure 6-9. (Anti-siphon valves are also known as *auto vents* or as *air admittance* valves). For details on installation, take a look at Art Ludwig's materials.

Laundry-to-landscape systems can also be installed if your washing machine is in your basement. If it's a walkout basement, you can simply run the pipe through a hole drilled in the basement wall. The washing machine pump and gravity are then generally sufficient to transport water to and through the irrigation field. In such cases, the design can be pretty simple.

If the basement is entirely below grade, however, you'll need to pump the water up and out of the basement, then let it flow by gravity into and through the drain field. Because washing machine pumps generally cannot pump water more than 3 feet (0.9 meters)

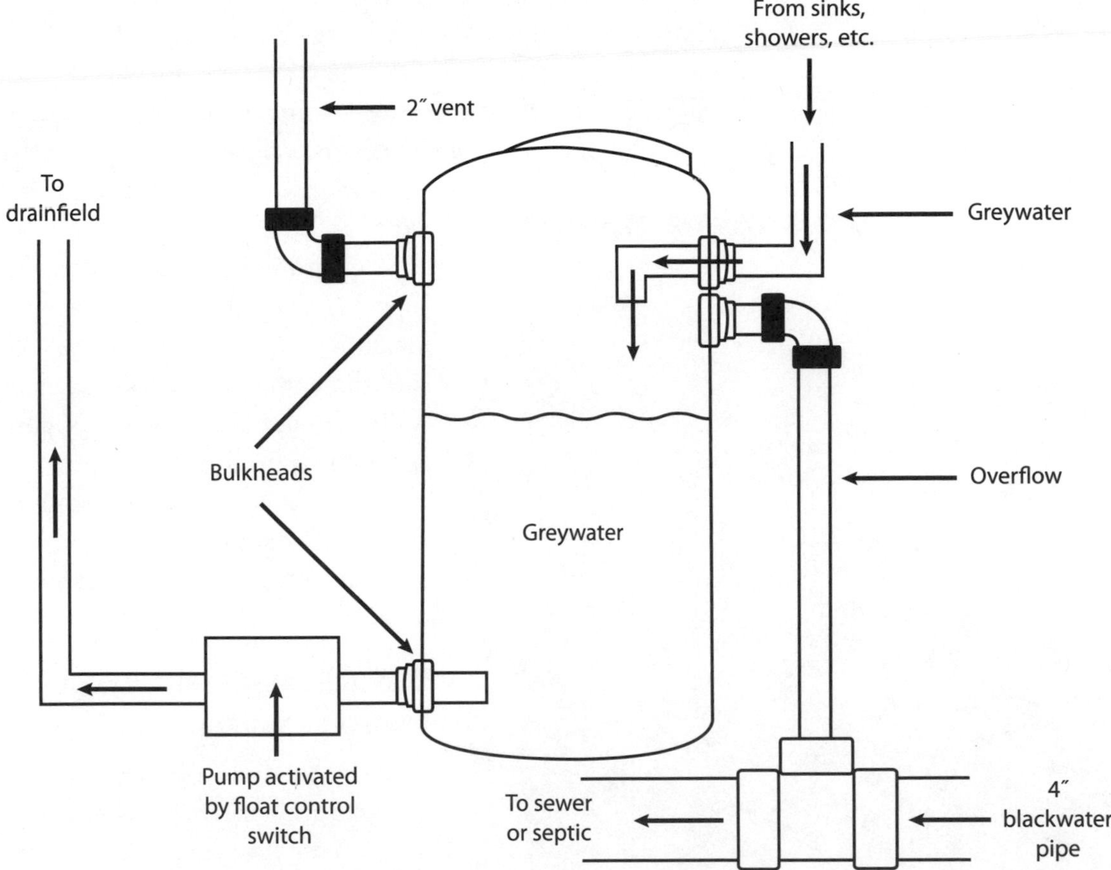

Fig. 6-10: Surge Tank and Pump. *A surge tank equipped with an external pump like this one next to a washing machine (not shown here) in a basement is needed to capture and then pump water out of the basement to the drain field.*

higher than the washing machine, you'll most likely have to empty the washing machine into a surge tank and pump the water out of it using a separate pump. For best results, install a pump *outside* the surge tank (Figure 6-10). The pump should not be immersed in the tank! Submersible pumps will not remove all of the greywater in the tank. And, as noted earlier in the chapter, greywater goes sour on you very quickly and will emit a foul odor.

One of the biggest challenges you will face when installing a laundry-to-landscape system is penetrating exterior walls of your home. Be sure to avoid damaging existing plumbing or electrical wires when cutting holes in framed walls. And, be sure you don't drill into studs. When drilling through a concrete wall, you may hit rebar, the steel reinforcing bars that are installed in concrete. No matter how hard you try, you can't drill through rebar with a concrete drill bit.

If you are uncertain about how to drill a hole in the side of your house or through your basement wall, hire an experienced professional or enlist the aid of an experienced friend. Also, be sure to seal up the penetration after the pipe has been installed to prevent air infiltration and water leakage into the home. Seal holes in framed walls with liquid foam spray (Figure 6-11). This product seals and insulates openings. Excess foam can be trimmed with a knife, then sanded, and painted. For concrete walls, silicone works well.

Fig. 6-11: Spray Foam. *This product helps seal openings in walls you drill for pipes. Be sure to seal all penetrations to keep insects out and prevent heat loss in the winter and heat gain in the summer.*

Pros and Cons of Laundry-to-landscape Systems

Laundry-to-landscape systems are low-cost and relatively easy to install — if you do it yourself. They require very little, if any, maintenance. And, of course, they allow you to irrigate your plants with water that would otherwise be wasted. If you design your mulch beds properly, there's little, if any, chance of contact with greywater. These systems are also flexible — that is, they can be altered if needs change. One big advantage is that they require no changes to existing plumbing.

On the downside, you will need to penetrate an exterior wall. And,

Cleaning the Filter in Your Washing Machine

All washing machine manufacturers install a filter upstream from the electric pump. Filters prevent pumps from getting clogged by fibers from clothing and damaged by coins, paper clips, and other junk from our pants pockets. It's a good idea to check the pump filter in your washing machine *before* starting up your greywater system just to be sure that it is free of debris.

Unfortunately, checking a pump filter can be quite difficult for the novice. Water pumps are generally accessed by removing a metal panel in the back of the washing machine. Unfortunately, manufacturers often place the pumps in nearly impossible-to-reach locations — beneath the drum toward the front of the machines — that offer insufficient room to work. The job begs for more hands that you were born with — and it's really difficult to find room for an assistant to reach in to lend a hand! You may want to hire a plumber to check and empty the filter. Maybe he or she will teach you how so next time you can tackle the job yourself.

you may need to acquire a permit to install a greywater system. So, be sure to determine whether you need a permit if you want to do this legally. In progressive greywater states, such as Arizona and New Mexico, no permits are required so long as you follow published guidelines.

As a final note, I'd be remiss if I didn't reiterate the importance of viewing Art Ludwig's excellent instructional DVD "Laundry to Landscape." It provides detailed coverage of this and only this system. I would recommend viewing it a couple times before attempting to install a system.

Now that you understand this option, let's look at another.

Branched Drain Systems

Figure 6-12 illustrates a greywater system known as a branched drain system. Also developed by Art Ludwig, it allows a homeowner to divert water from the rest of the house — that is, bathroom sinks, bathtubs, and showers. I'll give you a brief overview of the system, but as usual, complete, up-to-date information

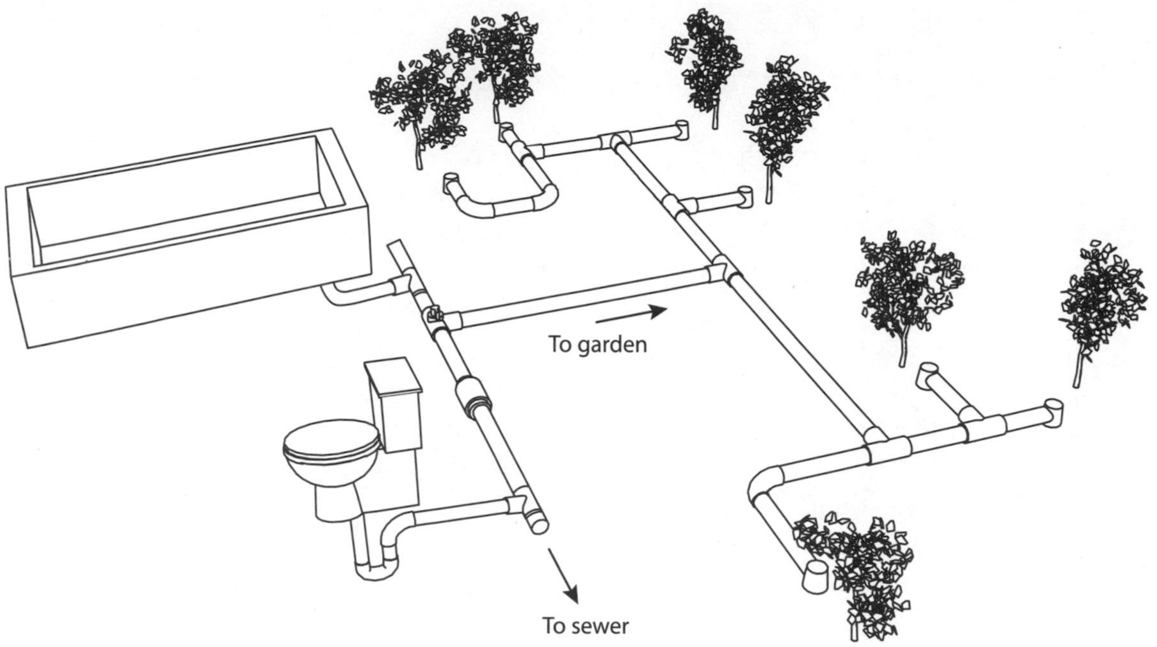

Fig. 6-12: Branched Drain System. *This system collects greywater from multiple sources in a home and delivers it to a fairly extensive drain field.* SOURCE: *CITY OF SAN FRANCISCO'S GREYWATER DESIGN MANUAL FOR OUTDOOR IRRIGATION*, P. 28.

on designing, building, and using this system can be found in Art's book *Create an Oasis with Greywater.* Be sure you check it out before installing this system.

A branched drain system requires a three-way diverter valve, as illustrated in Figure 6-1. As you recall, it allows a homeowner to divert greywater into the sewer/septic, for reasons you well understand by now.

Branched drain systems may be required to have a backwater valve, also shown in Figure 6-1. Backwater valves are installed in the drain pipe that connects sinks, showers, and tubs to the toilet drain. This valve prevents blackwater from mixing with greywater when the greywater irrigation system is operation. Should the need arise to divert greywater to the septic or sewer system, the diversion valve can be reopened.

When installing a three-way diverter valve, consider installing an actuator to control it. Actuators are electronic devices that allows one to remotely open and close diverter valves. Actuators are advised when the three-way valves are difficult to reach or if an operator simply wants a more convenient way of diverting greywater into the sewer/septic system.

Branched drain systems usually rely entirely on gravity to transport greywater out of the house and through the drain field. Irrigated plants must therefore be located lower than the source of greywater and the distribution pipes must slope ¼ inch per foot, or about 2%. No pumps or electricity is required, unless you install an actuator. A slightly larger (usually 1.5 to 2 inch or 3.8 to 5 cm) black ABS pipe is required in the landscape so that this pipe matches the size of the drain pipe used indoors. (Because my copyeditor wanted to know: ABS stands for acrylonitrile butadiene styrene. It's a type of plastic.)

Installing Two Greywater Systems

Many homeowners have found it advantageous to install two or more greywater diversion systems in their homes. For example, a shower and the sinks on one side of the house could be diverted to landscaping that's closest to them. A second system could be plumbed on the other side of the house. A laundry-to-landscape system could be a third system that irrigates plants in the yard closest to the washing machine. Clearly, it just depends on the layout of your home and landscaping.

Subsystems can have one enormous advantage over one large system: they help maintain "fall" — the amount of vertical distance between greywater sources and the drain field. Maintaining adequate fall helps ensure that water flows by gravity.

The main challenge you'll encounter is tying into the sink and tub drain lines, then running a greywater line outside your home. For single-story homes with basements, these tasks are typically accomplished in the basement — often in the spaces between the floor joists that are easy to access unless a ceiling has been installed.

Branched drain systems in existing two-story homes can be difficult to install because pipes joining the toilet drain pipe and sink or tub drains are hidden in floors and therefore inaccessible. (To get to them, you'd have to tear out a ceiling.) In such instances, a greywater system will require more work — and higher costs — though, it can be done. Because branched drain greywater systems require you to alter existing plumbing, you'll very likely need to obtain a permit from your city or county building department. Unless you're pretty experienced, it's probably a good idea to hire a plumber to install the inside pipe runs. Be sure they are given drawings like those in Ludwig's book.

If you are building a new house and want to incorporate a greywater system, be sure you or your plumber takes necessary steps to allow you to tap into greywater while the house is under construction. It's a lot cheaper to plump a greywater system while a home is being built.

Pros and Cons of Branched Drain Systems

Branched drain systems allow homeowner to make use of a lot more greywater than laundry systems do. That means that more plants can be watered, your water bills might be lower, and more water will be saved, benefitting the environment even more. For cities and towns, these systems help save energy, reduce water treatment, and reduce operating costs at water treatment plants. Widespread adoption of such systems could prevent costly expansion of existing water supply systems. And they could help protect water sources — aquifers and surface waters.

On the downside, branched drain systems are considerably more complicated, more difficult, and more expensive to install. And, they may require a permit.

Pumped Systems

In some instances, it may be necessary to install a pump to move greywater to a location upslope from the source of the greywater. "If this is the case, carefully review your situation to determine if it is worth installing a greywater system at all," advises Ludwig.

These systems require a surge tank and a water pump. The pump is activated by a float control switch. It turns the pump on when greywater starts filling the tank and shuts off when the tank is emptied. The surge tank needs to be vented, to release noxious odors. You also need to install an automatic overflow in case the pump fails and the tank overfills.

Some people install pumps to distribute water in drip irrigation systems. Drip irrigation systems are therefore pressurized. As such, they must include a pressure tank that maintains a constant pressure in the lines at all times, much like the pressure tanks for well water in rural homes supplied by wells. Because greywater is stored in the pressure tank, it needs to be filtered first to prevent it from becoming stinky and to remove gunk that can gum up irrigation pipes. For more details on this type of system, consult Art Ludwig's publications.

Installing a Drain Field

Installing a drain field requires a lot of careful planning. Stake out the field carefully, then mark your runs with spray paint or stakes. Once the drain field is laid out, measure run length to determine how much pipe you'll need to purchase. It's a good idea to draw the system as well and use that drawing to determine the type and estimate the number of fittings you'll need. After you've secured materials — including extra fittings and pipe — it's time to dig the ditch or lay the pipe on the surface.

For surface-run pipe, be sure to secure the pipe along its run in multiple locations. Pipe can be attached to the house and to fence posts. You can also install small stakes to secure the pipe. Also be sure to bury pipe in pathways or in mowed areas to prevent the hose being damaged by your lawn mower or people tripping over it.

If you're going to bury the pipe, you can use a trencher or dig the shallow runs with a shovel. Digging in the spring when the soil is moist always makes this task easier.

Digging ditches and laying pipe is pretty straightforward if your lawn gently slopes away from the house and the plants you want to irrigate are located along this slightly sloping landscape. Most yards are graded so that the surface of the lawn slightly slopes away from the foundation. This helps rain and snowmelt drain away from the house so it doesn't flood basements. If that's the case, dig a 4-inch deep trench throughout the pipe run. Water should flow nicely from the house to the irrigated plants.

If the land around your home does not slope away from the foundation or you are running pipe across flat areas, be sure to slope the ditch throughout its length. To maintain proper flow, the ditch needs to drop ¼ inch per foot. That may not sound like much, but it adds up quickly. Every 4 feet, the ditch must drop one inch. If you have to run the pipe 24 feet (7.3 meters), a ditch that starts out at 4 inches deep (10 cm) will be 10 inches (25.4 cm) deep at its termination. In 100 feet (30.5 meters), it would need to be 29 inches (74 cm) deep. Needless to say, long runs can be problematic, as they terminate below the mulch bed. In such cases, you maybe be able to eliminate or reduce the slope and rely on the pump to propel greywater. But pumps can only pump so far — so you will very likely be limited to about 100 feet of flat run.

To ensure proper slope, use a 4-foot level and a tape measure to check the depth and slope of the ditch as you dig. How do you determine slope with a level?

As shown in Figure 6-13, there are three lines on each side of the glass tube that houses the bubble. When the end of the bubble

Fig. 6-13: Bubble Level. *When a bubble settles between the inner lines an object is level. Obviously, the closer to center of this zone, the better. When the bubble extends into the next zones, on either side of the central region, the slope of the pipe is about ¼ inch per foot. Not all levels have this feature.*

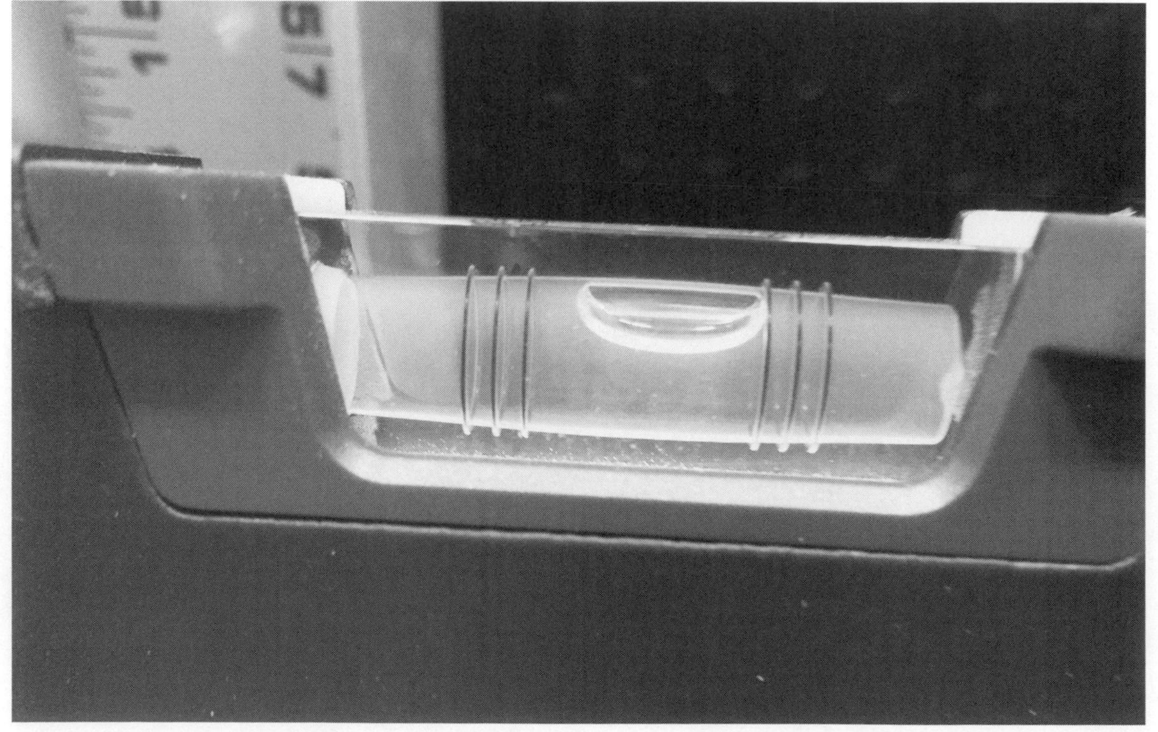

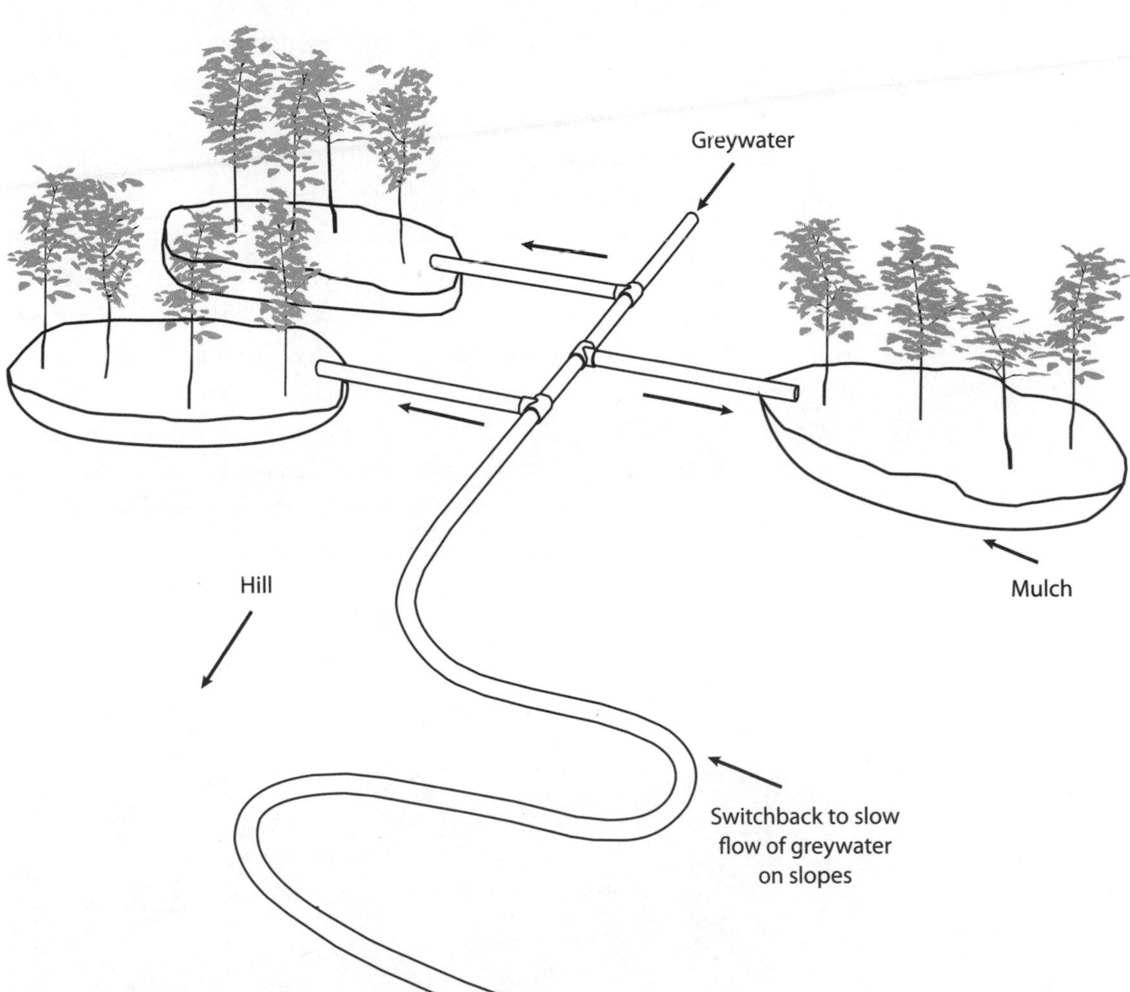

Fig. 6-14: Switchback. *Running greywater down hills can be tricky. Water tends to flow past upper emitters. To avoid this problem, install pipe in a switchback pattern and consider reducing pipe size in this section.*

is nestled between the first and second lines, the slope is ¼ inch per foot.

When installing distribution pipe on steep slopes, be sure to run it in a serpentine fashion — creating switchbacks — to control flow (Figure 6-14). Or, you can transition to smaller diameter pipe in the steep sections to control the flow. If you don't take such measures, greywater will rush to the bottom of the hill bypassing plants higher up in the irrigation system. (For more details on these systems, see the books and DVDs I've mentioned.)

Flexible tubing that you will run throughout most of the field requires barbed fittings. You can secure them with hose clamps if you'd like. However, it's not usually necessary to clamp elbows and tees to sections of flexible HDPE pipe because the pressure in these pipes is quite low and leakage, if any, will be minor.

Once you are happy with the layout, it is time to glue and/or clamp the pipe to the fittings. I cut pipe with an electric miter saw or conduit cutter. Both are much easier to use than a hacksaw.

Once the drain field pipes are installed, be sure to test the system — before filling in the ditches. Testing is vital to ensure that water flows to all the plants according to their needs. Testing is typically performed with fresh water delivered to the system from a garden hose attached to an outdoor spigot. A garden hose can be temporarily attached to the outdoor pipe as shown in Figure 6-15.

A better option is to install a permanent hose fitting at one location in the flexible pipe. It will allow you to attach a garden hose whenever you want to flush the system. The connection should be near the beginning of the outdoor pipe run. The faucet to which

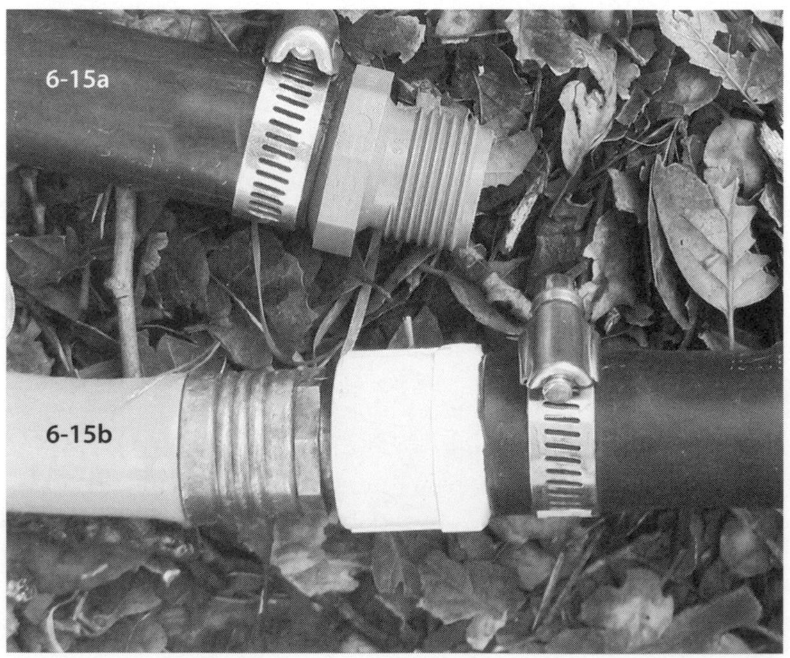

Fig. 6-15a and b: Fittings Required to Attach a Hose to Black Pipe. *In 6-15b, a garden hose is temporarily attached to the black one-inch HDPE pipe.* Courtesy of Art Ludwig

Fig. 6-16: Backflow Valve. *Any time a greywater system is connected to a hose attached to a faucet , a backflow valve is required to prevent greywater from entering the pipes that deliver water to fixtures in the house.*

you attach the garden hose should be equipped with a backflow valve to prevent greywater from entering the household water (Figure 6-16). Back flow valves are screwed directly on to the hose bib (outdoor faucet).

Using a garden hose to test the system will ensure a steady supply of water and adequate time to check for leaks and observe and adjust flow rates. For a laundry-to-landscape system, you may want to hook the system up to the washing machine first to determine how quickly water flows out of the appliance. Then match the flow rate with the garden hose.

As you will soon find, water will very likely flow out of the first openings in the drain field more quickly than from downstream openings. If this occurs, you will need to install some means of reducing outflow early in the pipe run so that plants at the end of the run receive sufficient water.

Installers employ one of several measures to control the flow at each mulch basin. One way is to step the pipe down at each mulch basin — that is, reduce its diameter.

A laundry-to-landscape system, for example, one-inch HDPE flexible plastic pipe typically runs throughout the drain field.

However, at each plant high in the system, the one-inch flexible tubing terminates on ½-inch pipe. It runs to the mulch shield, described earlier.

With the garden hose supplying a steady stream of water that mimics the washing machine, walk down the line to observe water flow at each plant. If it is flowing too quickly, adjust the connection of the one inch to ½-inch pipe (a tee fitting) so the ½-inch (1.25 cm) pipe runs uphill a little. This simple measure slows the water release at that point. If this technique is not sufficient to slow the flow, insert a tiny valve in the end of the pipe that empties into the mulch basin. These valves can be adjusted to fine tune flow rates. Experts recommend ½-inch (1.25 cm) green (green back valves) or similar size purple ball valve. (These valves tend to resist clogging, so only use these types.) Cranking the valve down reduces the flow of water. If water is flowing too quickly out of the next outlet, adjust the tee or install a valve on that one. Continue down the line until water flow to each plant is perfect. Please note, even though the green or purple ball valves are pretty reliable, they will clog. So, avoid adding too many.

If you are relying on your washing machine pump to help dispense water through the drain field, be sure that that end of the one-inch trunk line that runs through the entire system is not capped off or regulated by a valve. If capped, the pipe could create back pressure that could damage the water pump.

After testing flow rates with a garden hose, attach the drain field to the washing machine and run a load of wash. As the washing machine drains, check for leaks at all fittings inside and outside the house. Then check flow rate at each outlet. You may need to fine tune the system a little as the water pressure in the laundry hose may differ slightly from that in the garden hose.

With a branched drain system draining sinks and showers, be sure to run tests while two or three fixtures are draining to simulate real-life conditions. For instance, you may want to test the system when the bathtub is draining and the water is running in a shower. Or if you prefer showers, test the system with shower water and a sink or two running.

Once the flow rates have been adjusted, take photographs of the drain pipe for future reference — in case you have to dig them up or run other pipe or wire underground. You can now safely backfill buried pipe. When you do, be sure to mound the dirt over the excavations and then compress it with your foot. The mounded dirt will settle over time even if you have compressed it.

It's often a good idea to reseed the disturbed area. The fastest way to reseed an area is to apply grass seed over the bare dirt, then rake it in. I cover the area with loose straw and water every day or so for a couple of weeks. Straw helps keep the moisture in and will promote rapid growth of the grass.

If you have a lot of plants to irrigate, you may want to divide your system into two or more zones. One zone can be watered one week, the second zone the next week, and so on. To control the flow to each zone, you'll need to install underground three-way brass valves. Install them in buried valve boxes in readily accessible locations. Because brass three-way valves are fairly expensive, you may want to install two ball valves, one on each branch. Opening one and closing the other will divert water to the zone to which you want it to flow (Figure 6-17). Another option is to design the

Fig. 6-17: Ball Valves. *Ball valves are fairly inexpensive and can be teamed up to avoid having to install a more expensive three-way valve.*

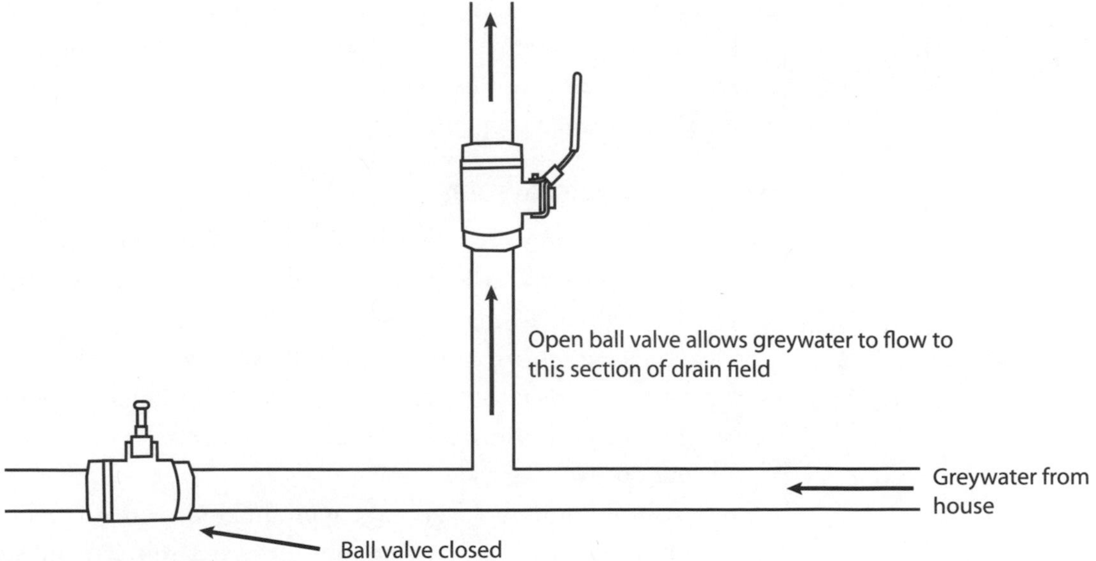

greywater irrigation system so that one zone is supplied by the washing machine and the other is supplied by sinks and showers.

As a final note, if you have to pump water uphill, do so carefully. Consult with Art Ludwig's book. Be sure to install proper backflow valves to prevent greywater from flowing back into the washing machine when the pump turns off.

Labeling Greywater Systems

After laying pipe, be sure to label all of the lines and valves in the system. Labeling will help you remember where each pipe runs. It is information that will be valuable should the system need servicing. It will also will help others — like family members or a future homeowner — operate and maintain the system.

Start by labeling the three-way diverter valve in your laundry-to-landscape system. Place a label on the valve on the wall. Label the pipe as well to indicate which pipe delivers greywater to the septic/sewer (an ocean or river) and which one delivers it to your plants. Be sure you affix labels that will last forever and use indelible ink. No printer labels with water-based inks, please!

Be sure to label greywater lines outdoors, too, especially near the house. You may want to label at several locations along the irrigation field. Label outside diverter valves, too. It's a good idea to sketch a map of the system and prepare instructions on proper operation. You can give this to the next homeowner or to a renter or to your kids when they inherit your home.

Sizing a Greywater System

The design of a greywater system requires an estimate of the amount of greywater a home produces and how much water plants require. Unfortunately, that's much easier said than done.

Irrigation requirements vary considerably. In warm desert climates, you may irrigate several times per week throughout most of the year. In colder climates, you may only irrigate once or twice a week and only during the summer.

To determine household greywater production, you can estimate water production by each of your greywater sources: sinks, showers, tubs, dishwashers, and clothes washers. Art Ludwig provides a handy table with estimates of water production from each source in his book *Create an Oasis with Greywater.*

If your home is supplied by a municipal water utility, you can estimate greywater production using your monthly water bill. First, determine the average daily water demand during that part of the year when you are *not* irrigating your lawn, gardens, and trees. Add up the gallons "consumed" and then divide by the number of months.

To determine the amount of greywater your family produces, multiply the total monthly water consumption during each month when no irrigation is required by 0.75. That will give you a fairly good estimate of monthly greywater production. For example, suppose your family consumes 12,000 gallons of water per month in the winter. Greywater production would be about 0.75 times 12,000 gallons which equals 9,000 gallons per month. Now, how do you calculate irrigation demands?

Let's suppose that your family consumes 24,000 gallons of water per month on average during the summer. To determine irrigation water requirements, subtract the average monthly water demand in the winter from the average monthly summer use. The difference will be the monthly irrigation requirement — that is, how many gallons of water are required to irrigate your lawn, gardens and trees each month. If you subtract 12,000 gallons from 24,000 gallons, you will see that the house uses about 12,000 gallons of

How Many Plants Will a Washing Machine Irrigate?

According to the City of San Francisco's *Greywater Design Manual for Outdoor Irrigation,* a front-end loading washing machine can typically supply eight mulch basins. A top-loading washer, which uses more water, can supply about 12 mulch beds. (Consider these as general guidelines for San Francisco and be sure to run the math on your own system.) Art Ludwig says you can water one to two medium fruit trees per person from the washer water. He lives in Santa Barbara in southern California where it is considerably drier. To fine tune your system, ask a local nursery how much water each tree or shrub needs in your region, then determine how many gallons each load of laundry produces and how much additional greywater you will produce elsewhere in the system.

irrigation water per month. So, greywater would cover two-thirds of your demand.

Remember, this is an average. As you'd expect, water requirements may be higher in July and August, the hottest months, than June and September, usually cooler months.

Remember also if you have a swimming pool, average monthly irrigation demands will very likely be lower than your calculations show.

Also bear in mind that these numbers include lawn watering. Determining how much water is used for gardens versus trees versus the lawn is difficult. Interestingly, most summer irrigation water is dedicated to lawn watering.

Ultimately, you'll have to estimate the amount of water needed to irrigate your lawn versus gardens, trees, and shrubs based on experience. If you know how often you irrigate each lawn or each tree, bush, and garden and can determine the flow rate in the hose, you can estimate how many gallons of water it takes to irrigate each one and estimate demand for each one.

You can also estimate the amount of water each tree or shrub requires by asking an experienced arborist, irrigation professional, or knowledgeable employee at a local nursery. A master gardener may have some general guidelines for watering gardens.

Lest we forget, you'll also need to factor in rainfall. The wetter the climate, the less greywater you will need. The drier the climate, the thirstier the plant, the more greywater you will need to provide.

Whatever you do, don't drive yourself crazy trying to determine exactly how much water you'll need. "Estimating irrigation demand is an inexact science," notes Art Ludwig. "Even getting within a factor of two of real irrigation demand is an ambitious goal." To ease your mind, Ludwig also notes that "the vast majority of greywater systems are made without any calculations at all — and most still work." So don't lose any sleep over this. Do your best and monitor your plants throughout the irrigation season. When dry conditions persist or plants are looking a bit droopy, supplement greywater with rooftop rainwater or city or well water.

Installing and Sizing Mulch Beds

Mulch basins need to be excavated around every plant you are going to irrigate. As a friendly reminder, mulch basins are installed to ensure that greywater quickly enters the soil so potential contaminants can be rapidly biodegraded into useful nutrients and water is taken up by roots. Dig mulch basins before you lay the pipe.

Most mulch beds circle the plants they serve like moats. If that's not possible, a semicircular trench will very likely suffice. Dig the basin 6 to 12 inches (15 to 30 cm) deep. Remove the topsoil. Use some of the soil to create a mound around the mulch basin to keep rainwater from flowing in (Figure 6-18). This gives you more exact control over watering. For a detailed look at mulch basin construction, check out Ludwig's books. They contain many excellent drawings.

Fig. 6-18: Mulch Basin and Root Island.

After successful excavation, fill the moat with mulch, then install the mulch shield and pipe. Mulch should be fairly coarse to allow water to flow rapidly through it and into the underlying soil. Wood chips make great mulch.

You may be able to obtain wood chips from a local tree trimmer. They will probably give them to you and might even drop them off at your place for free if it's not too far out of their way.

Mulch can also be obtained from cities and towns. They often chip trees or limbs that they cut down in parks and along roadways. Some add the chips to organic matter (lawn trimmings, etc.) as part of city-wide composting activities. Don't use this as mulch.

If you are really ambitious, you can make your own mulch. You can rent or borrow a chipper from a friend or neighbor to grind up tree limbs. You might ask neighbors to save tree trimmings for you. Stockpile them so when you have a decent pile you can bring in the chipper, but watch your fingers!

If neither source is available and you're not up to making your own wood chips, you can purchase mulch. Home centers and nurseries sell mulch by the bag or by the ton. It's best to use undyed products.

Mulch beds are "installed" from the trunk to the drip line of plants — that's the outer perimeter of the canopy (the leafy portion of the tree or bush), as shown in Figure 6-19. Because the drip line will shift as a tree or bush grows, you will need to extend the mulch bed over time. That said, as trees grow larger and their root systems become more extensive, irrigation may no longer be needed except in extreme drought or when nearby excavation disturbs the root zone.

Mulch bed design depends on a plant's water preferences. For plants that need to periodically dry out, be sure the root crown forms an island in the moat, as shown in Figure 6-19. This section will dry out between waterings.

Plants like bananas, on the other hand, can tolerate and may even prefer their roots be immersed in moist soils. Ask your local nursery or a qualified arborist for root moisture preferences of

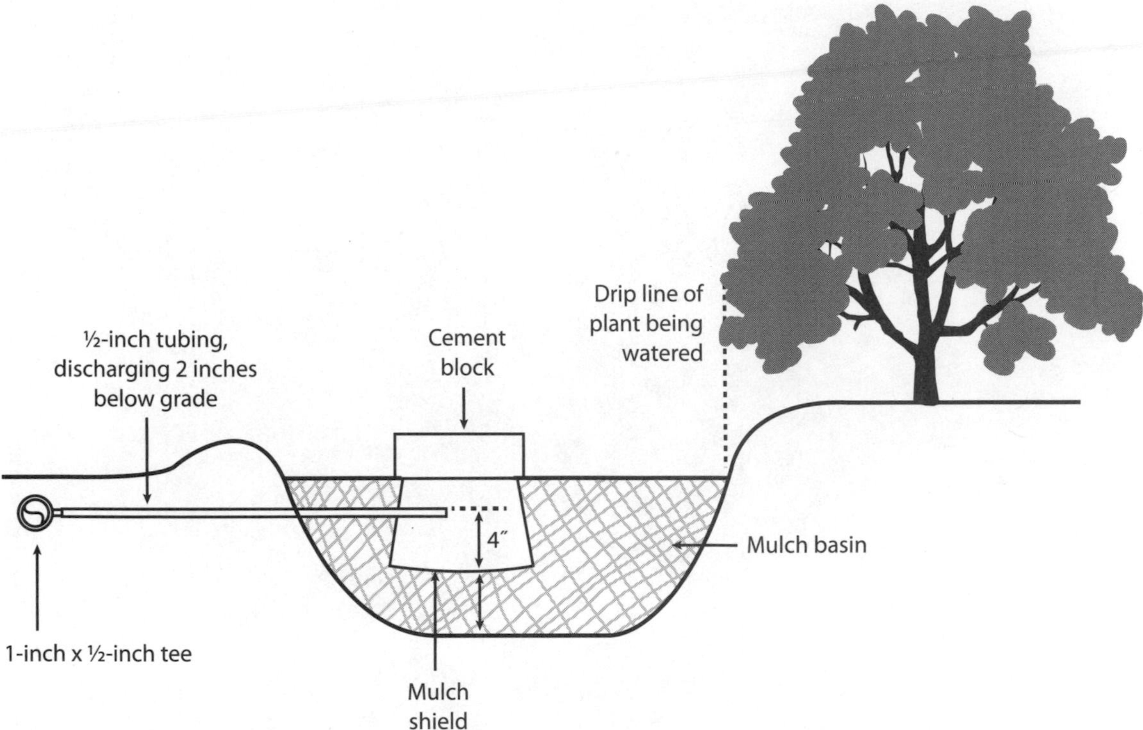

Fig. 6-19: Mulch Shield. *Mulch shields can be made out of plastic flowers.*

the trees and shrubs you intend to irrigate. I'd highly recommend consulting with Ludwig on this topic. Money spent on consulting will be well worth it!

Mulch shields can be made from one- to three-gallon (4 to 12 liter) plastic flower pots. To make a mulch shield, turn the pot upside down, and drill a hole in the side for the pipe at the right depth (Figure 6-19). If the pipe enters at four inches, that's where the hole should be drilled.

Using a sharp utility knife or a stout pair of scissors like Fiskars, cut off the bottom of the pot to create an operable lid. Don't cut it all the way off. Leave enough material to serve as a hinge. It's not a bad idea to leave three or four inches of plastic for a hinge. The hinged lid allows you to inspect the outlet from time to time.

The bottom of the inverted pot should extend about 4 inches (10 cm) below the entry point of the drain pipe. Trim off excess plastic to achieve this depth.

Fig. 6-20: Valve Box. *Valve boxes can be used to protect emitters and valves in a greywater system.*

If you are not up to making your own mulch shields, you can use valve boxes. They can be purchased from hardware and home improvement centers and modified as needed (Figure 6-20). They're found in the plumbing section.

What Products Can I Use in My Greywater System?

Now that you have an idea how greywater irrigation systems are designed and how they work, let's examine some rules of operation that will ensure that your system works well. We'll focus on safe chemical agents you can use to wash clothes, shower, and so on.

Unfortunately, most people in more developed countries like Canada and United States pour tons of toxic cleaning agents and solvents down their drains with little, if any, thought of the consequences. Successful greywater systems require the use of plant- and soil-safe cleaning agents.

I learned the importance of this early on in my greywater experimental years. Shortly after I started using greywater, I purchased some borax (sodium borate) to use in my washing machine — without performing any research on this compound. (My first mistake.) Sodium borate seemed like it would be a better "bleaching agent" than conventional bleach (sodium hypochlorite).

Unfortunately, I soon discovered that this was an erroneous assumption. Shortly after using this product, I noticed that the leaves on all my plants in my indoor planters started turning brown around the edges. Suspicious, I decided to do a little research — the investigation I should have performed upfront! That's when I learned that boron is toxic to plants. I stopped using the powdered laundry cleaning agent immediately and my plants quickly recovered.

To irrigate with greywater, you'll need to abandon all conventional cleaning agents and detergents — as most contain chemicals that will poison plants or your soil. First and foremost, be sure to avoid the toxic trio: chlorine bleach, soaps containing salt (sodium chloride), and products containing boron (like Borax). The latter two are nontoxic to people, but harmful to plants. What can you use instead?

My advice is to use nontoxic, biodegradable cleaning agents. But shop carefully. Not all biodegradable products are plant friendly. (More on this shortly.)

Second, avoid cleaning agents that alter the pH of greywater. Bar soaps, for instance, raise the pH of water, making it more alkaline. Blueberries and other acid-loving plants such as ferns, azaleas, camellias, and rhododendrons require acidic soils. They will not grow well in basic soils. Instead of bar soap, use liquid soap. They generally do not alter pH.

For clothes washing, I have used Oasis Laundry Detergent with great success for many years. This product, which as I understand was developed by Art Ludwig, is a specially formulated concentrate that contains biodegradable components that break down into plant nutrients (Figure 6-21). This product is not just biodegradable, it is considered biocompatible.

Fig. 6-21: Oasis Biocompatible Laundry Detergent. *I've used this laundry detergent in my greywater system for many years with great success!* Courtesy of Bio Pac Cleaning Products

The nonprofit group, Greywater Action, recommends the following laundry detergents for use in greywater systems: Oasis Laundry Detergent, ECOS liquid detergent, and Dr. Bronner's liquid soap. All of which can be purchased online.

Instead of bleach, give hydrogen peroxide a try. Many consider it a safer alternative to chlorine-based bleaches. I have never used this product, so can't attest to its efficacy or safety.

For washing dishes and your hands, Greywater Action recommends the following products: (1) Oasis Biocompatible Dishwash/All Purpose Cleaner, (2) Dr. Bronner's liquid soap, and (3) any natural liquid or bar soap. Aubrey Organics manufactures and sells shampoos and conditioners that don't contain salt or unhealthy chemicals.

Permits for a Greywater System

As noted earlier in the chapter, when installing a greywater system in a new or existing building, you may be required to obtain a permit from your local AHJ (authority having jurisdiction). That's your city, town, or county building department. Call them to determine what's required. You may even be required to apply for a permit through the state.

Whether you need a permit in some cases — for instance, in San Francisco — may depend on whether you alter existing plumbing. In California, laundry-to-landscape systems do not require permits, so long as the homeowners do not alter existing plumbing. The rest of the greywater systems require permits.

Building permits typically require a drawing of the system and may require a stamp from a local engineer. This could cost $300 to $1,000, depending on where you live and the sophistication of your system. The building department will also very likely charge a fee and will perform at least one inspection of the system, usually when it is completed.

If you live in states like New Mexico, Arizona, and Texas that have adopted rational requirements for greywater systems, consider yourself lucky. If you live in states still mired in the Dark Ages when it comes to greywater, you may be in for a bit of a slog.

Greywater system guidelines are outlined in an appendix in the Universal Plumbing Code (UPC), a set of regulations that some AHJs have adopted. It was based on the greywater appendix from the California Plumbing Code which Ludwig describes as "maddening and unreasonable." So maddening and unreasonable is it that most greywater systems in western states are unpermitted. That is to say, they were installed in new or existing homes "under the radar" or, as I like to say, "without the benefit of Code."

I can't recommend subverting Code, and installing a shoddy system is never a good idea. You should always install a system that is safe, effective, and properly sized. That said, Code-approved systems can be unnecessarily complex, costly, and may require a lot of maintenance. As if that's not bad enough, according to Ludwig, they often don't work very well, either. Filters clog and pumps burn out. Simple, homespun systems lacking pumps and filters are inexpensive, effective, and can last a long time. Imagine that!

Many jurisdictions are improving their greywater regulations. California now allows Art Ludwig's laundry-to-landscape systems without a permit, and Arizona and New Mexico and other states allow all greywater systems without a permit.

Permits will very likely be required for systems designed to service multi-unit buildings, businesses, and institutions. These systems are often sophisticated. Even though they are well engineered and fairly costly, they can be quite cost effective.

If you are building a new home, remember that it is almost always less expensive to install a greywater system when the home

"Legal requirements (for greywater systems) favor engineering overkill, while the simple and economical methods — the ones that people actually use — remain technically illegal."

— Art Ludwig, *Create an Oasis with Greywater*

is under construction than after construction is complete. That's because it is easier to install pipes and valves inside the house alongside conventional plumbing. It's also easier to bury pipes in the yard if it hasn't been seeded or sodded. In new construction, you'll very likely need to obtain a permit or fly under the radar, retrofitting the plumbing after the rough-in plumbing inspections. I can't recommend this strategy, but it's been done.

Pros and Cons of Greywater

By now, you have a pretty decent idea of the benefits of greywater. A quick recap may help crystallize your knowledge and help you make a decision to install a greywater system. Even a laundry-to-landscape system will help.

For those readers who are committed to self-sufficiency, greywater systems help us reach that goal, as they can be used to irrigate a wide range of edible plants — from fruit trees to edible flowers to berries to vegetables. It also helps us create a more beautiful landscape by supplying ornamental trees. Shade trees help keep our homes cooler in the summer, beautify our homes, reduce our carbon footprint, and save money.

Greywater can also be cleaned up and used to flush toilets and urinals. However, this is rarely done. Purified greywater can even be used to irrigate lawns with subsurface drip irrigation pipes. If you are living in the International Space Station, greywater can be cleaned so thoroughly that it can be used to shower and drink.

Utilizing greywater for these and other purposes reduces the amount of water that must be pumped out of aquifers and rivers, lakes, and streams. This, in turn, helps to protect the habitat of fish and other wildlife that depend on aquatic ecosystems. Purified by the topsoil, greywater can also help replenish groundwater supplies.

Greywater use can be an extremely valuable ally in water-short areas — regions suffering from chronic or acute drought. According to one source, greywater systems can reduce the consumption of potable water by 16% to 40%. As water shortages worsen, numerous states are allowing — and even encouraging — the use of greywater.

Reducing water demand saves money for home and business owners. Because sewer rates in cities and towns are typically based on the amount of water consumed by a household each month, you'll often pay less for sewage disposal as well.

Greywater use also helps to reduce the amount of sewage delivered to sewage treatment plants. This lowers energy use at these facilities. It also reduces the use of toxic chemicals, such as chlorine, which are added to treated sewage in many plants before it is released into streams or rivers. Because it takes energy to make and transport chlorine to sewage treatment plants, reductions in chlorine use helps us save energy and reduces air and water pollution.

Reducing the amount of greywater sent to sewage treatment plants and septic tanks also reduces strain on overtaxed or failing systems. Domestic greywater systems can also divert huge amounts of water from failing septic tank leach fields, saving costly replacement. In addition, greywater systems can be installed in areas where clay-rich soils deter percolation from leach fields. Rural Missouri is one such example.

Because greywater often contains small amounts of organic matter, its use for plant irrigation not only helps provide water but also some nutrients plants need to grow and bear fruit or flowers.

Greywater use is safe, too. Because greywater is deposited just below the surface or directly onto highly porous mulch beds, it won't pool up. It percolates into the topsoil surrounding the roots of plants. Topsoil is teeming with microorganisms that quickly gobble up any organic matter in greywater, releasing valuable nutrients plants need to grow and produce edible fruit, leaves, or seeds. To enhance greywater's value, special detergents can be used — adding more nutrients to the soil.

While greywater offers many advantages and can be used in many locations, there are some situations in which you may not want to install a system. These include: (1) homes with postage-stamp-sized lawns; (2) homes where the plumbing is inaccessible, for instance, it is installed beneath or in concrete slabs; (3) very wet regions with impermeable soils; (4) homes that require highly engineered and expensive systems to treat and pump greywater and; (5) areas where code officials don't approve of them (although code officials may turn a blind eye to them, according to greywater guru, Art Ludwig).

Do Greywater Systems Make Economic Sense?

Whether a greywater system makes economic sense ultimately boils down to three key factors: the initial cost, water cost, and how greywater is used.

The initial cost of greywater systems vary greatly, depending on their complexity. The simpler they are, the less plumbing you'll need, the lower the cost. Another price-determining factor is size. The bigger the system, the more costly. A third factor is who installs the system. When a professional designer or installer enters the picture, costs can rapidly escalate. As you may find, their labor costs can be significant. If you design and install the system yourself with some knowledgeable friends, you'll invariably pay a *lot* less. Keep in mind, in low-tech systems, most of the labor is in digging — excavating basins and trenches. If you can do that, enlist the aid of folks who want to learn, or hire some hard-working high school students, excavation costs can be dramatically reduced.

Water costs also affect the economics of a system. The more you pay for water, the more cost effective a greywater system becomes. Finally, how you use greywater will factor into its economics. Systems designed to provide irrigation water often make economic sense.

For homeowners and small businesses, low-tech greywater systems used to irrigate the landscape generally represent a good investment of time and money. According to Greywater Action, a simple laundry-to-landscape system will run you $100–$250 for materials. With professional installation, the system could cost from $700–$2,000. Branched drain systems will run you $200–$400 for materials, with full installation bumping the price up to $800–$3,000. For pumped systems, materials could easily run between $400–$1,000, depending on the type of pump that one installs. Professional installation could increase the price considerably, adding another $1,000 to $4,000.

If you install a really complex system with filters, expect to pay a lot more, in the range of $5,000–$10,000. Even so, in new construction, these more complex, engineer-designed and contractor-built systems may make economic sense, so long as the irrigation water savings equals or exceeds 200 gallons per day. That's not difficult to achieve in a household of four.

When designed to purify greywater for flushing toilets, greywater systems generally don't make economic sense. They are not only not economic, they may not be environmentally sound. More costly engineered systems often only make economic sense if they serve a larger population — for example, residents of an apartment complex, office complexes, and so on — and when the water is used to irrigate vegetation.

Conclusion

Greywater systems help put a wasted resource into use and have so many benefits that they ought to be installed in every home in the world! Although that's not going to happen, you can do your part in capturing and using this valuable resource while helping to create an oasis to live in. This chapter has provided the basics. If you are interested in pursuing this option, be sure to study more. There's a lot more to know. Equipped with the information I've shared with you, however, it should be pretty easy to gain the knowledge and skill you need to create a successful greywater system.

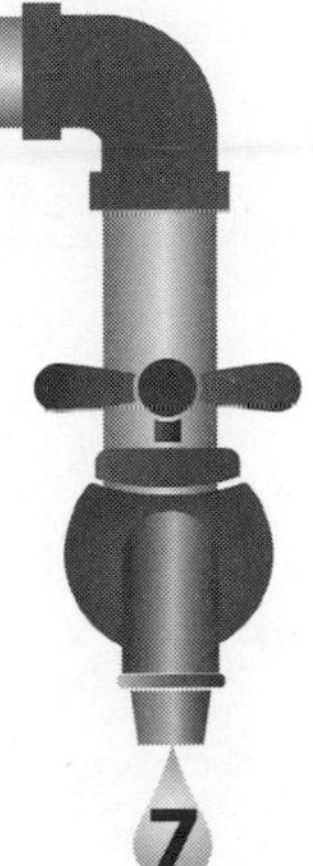

Residential-scale Constructed Wetlands

Biologists often liken tropical rainforests to the lungs of the planet. This analogy is based on the fact that the luxuriant vegetation of the rainforests produces massive quantities of oxygen via photosynthesis. Oxygen, of course, is vital to virtually all organisms on planet Earth. As apt as this analogy is, however, it is actually a half-truth — well, truthfully, a two-thirds truth. As it turns out, microscopic photosynthetic organisms like algae — known as phytoplankton — generate about one-third of the oxygen required by all animal life, including humans, on planet Earth.

If rainforests and phytoplankton are the lungs of the planet, wetlands are the planet's kidneys. Wetlands form along the banks of rivers, ponds, lakes, and estuaries, or they may form in depressions in fields and forests. Wetlands' role as the kidneys of planet Earth stems from their ability to filter out sediment and other harmful pollutants carried in waters flowing into and through them. Sediment is produced by erosion that results primarily from poorly managed farms and badly designed roads and construction sites. In filtering out sediment, wetlands help to purify water — and they do so far more efficiently and inexpensively than the best sewage treatment plants humans have ever devised.

Wetlands are also capable of detoxifying harmful chemicals and thus act a lot like our livers. In the world's many wetlands, waterborne pollutants are removed by microorganisms that make their home in its organic-rich muck and on the roots and stalks

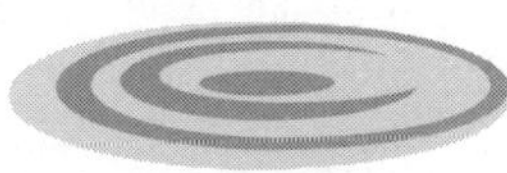

of wetland plants. These microorganisms decompose or assimilate all manner of organic and inorganic pollutants from many sources: factories, farm fields, sewage treatment plants, parking lots, and lawns. Wetlands can even remove and neutralize fairly toxic chemicals.

Like compost piles containing humanure, wetlands also remove and destroy a number of pathogens — that is, potentially harmful microorganisms from human and animal feces. In this capacity, then, wetlands act like our immune systems.

Because wetlands typically consist of a dense maze of aquatic plants, water flows slowly through them. This allows plenty of time for the removal and decomposition of pollutants and potentially harmful microorganisms.

Wetlands, then, are the planet's kidneys, liver, and immune system. Unfortunately, in our race to get ahead, many countries have destroyed valuable riverine and coastal wetlands. They've largely been filled in (we erroneously call it "reclaimed") to make room for cities, towns, subdivisions, farms, factories, and power plants. To give you an idea of the scope of this transgression, US wetlands once covered an area twice the size of California. At least half of all these wetlands have been destroyed. In some states, like California, Iowa, and Ohio, around 90% of the wetlands have vanished at the hand of humankind. Figure 7-1 shows a map that indicates the percentage of wetlands that have fallen victim to human development.

Because wetlands are so effective in purifying water, many cities and towns throughout the world have begun to construct artificial wetlands to treat human sewage. These are referred to as *constructed wetlands.* By various estimates, in the United States there are about 1,000 constructed wetlands in operation today. Europe boasts five times as many.

If carefully designed and maintained, over long periods they come to support a rich and diverse array

Wetland Decline

Many nations have destroyed large percentages of their wetlands to make room for cities, towns, farms, and factories. US wetlands once covered an area twice the size of California. At least half of all these wetlands have been destroyed. In some states, like California, Iowa, and Ohio, around 90% of the wetlands have vanished at the hand of humankind.

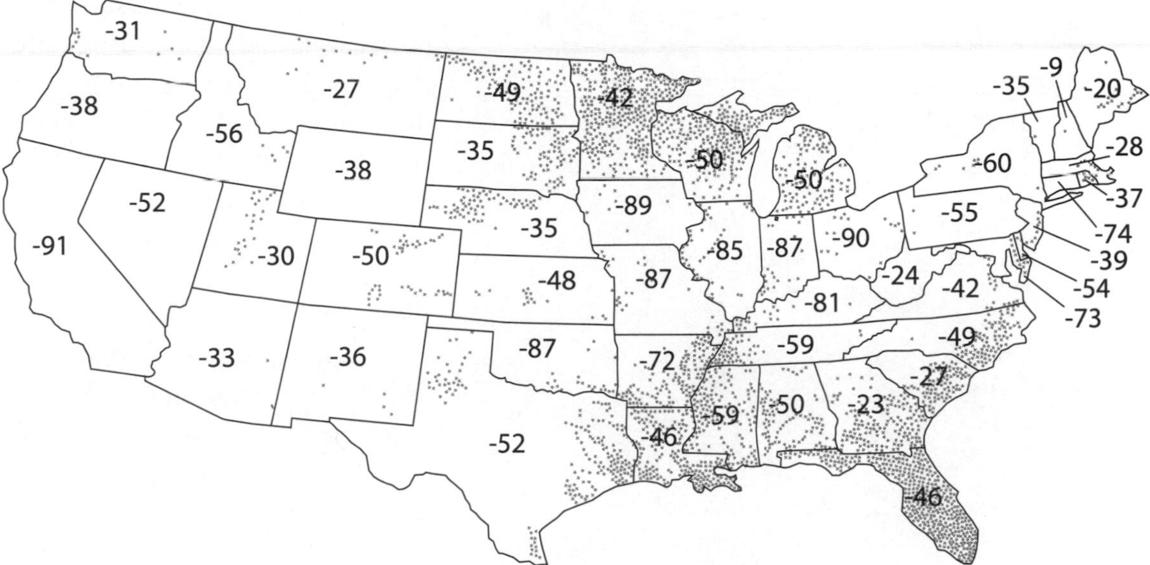

Fig. 7-1: US Wetland Loss. *Wetlands are the kidneys, livers, and immune system of the planet, but have been destroyed by actions taken before their true value was understood.*

of plant and animal species. These "artificial" biological water purifications systems can become fully functioning wetlands. Short on open space to build wetlands, some communities treat their sewage in large shallow, open concrete ponds or tanks. In them, they grow natural water purifiers — plants such as duckweed and water hyacinth. These species absorb organic and inorganic nutrients in sewage, and thus help to purify it. Both duckweed and water hyacinth grow extremely rapidly and therefore must be periodically harvested. If free of toxic pollutants, though, duckweed can be fed to ducks, chickens, and livestock.

Duckweed is a protein-rich food source. Amazingly, a duckweed pond produces six times the protein of a soybean field (per acre). In less developed countries, duckweed is routinely harvested from ponds and fed to animals. In 2014, I started growing duckweed in 275-gallon totes fed by partially filtered water from our duck's swimming pool. Water from the pool first flows into a biological filter that consists of scrubber pads suspended in a mesh minnow bucket suspended in a 55-gallon plastic drum. This filter removes some of the organic matter and most of the algae (Figure 7-2). The filtered water then flows into other totes in which we grow

Constructed Wetlands

Constructed or artificial wetlands mimic natural wetlands. They support aquatic plants and animals and support a number of naturally occurring biological and physical processes that remove pollutants from water.

Fig. 7-2: Filter. *The author's first duckaponics system in operation requires a filter made from scrub pads suspended in a 55-gallon plastic barrel.*

lettuce, cucumbers, and squash in pots suspended on floating rafts (Figure 7-3a and b). Several tanks in the system are dedicated to growing algae and duckweed, both of which we feed to our ducks.

Water purification in constructed wetlands requires little, if any, energy and no harmful chemicals. It operates quietly and efficiently without producing noise or odor. It represents some of humankind's best efforts to design systems with nature. This strategy helps us "fit" human systems into the natural systems — the Earth's ecosystems, which — and we often forget this — are the life-support system of the planet.

The vast majority of constructed wetlands are designed to serve municipal sewage treatment plants. Municipal constructed wetlands receive partially treated sewage. The effluent is pumped into the wetland for "polishing" — final purification — before being released into waterways. Chances are you have seen a constructed wetland and not even known what it was.

Constructed wetlands can also be used to capture many types of undesirables: pollutants from livestock and poultry operations; runoff from streets, parking lots, and lawns; and even organic waste from paper mills. More to the point of this chapter, some homeowners throughout the world are building small constructed wetlands to capture water and nutrients from their homes and return them to nature. In this chapter, we'll focus our attention on

Fig. 7-3: Duckaponics System. *(a) Water flows from the barrel to four totes used to grow food for his family and his ducks. (b) Young squash growing in net pots suspended in duck water.*

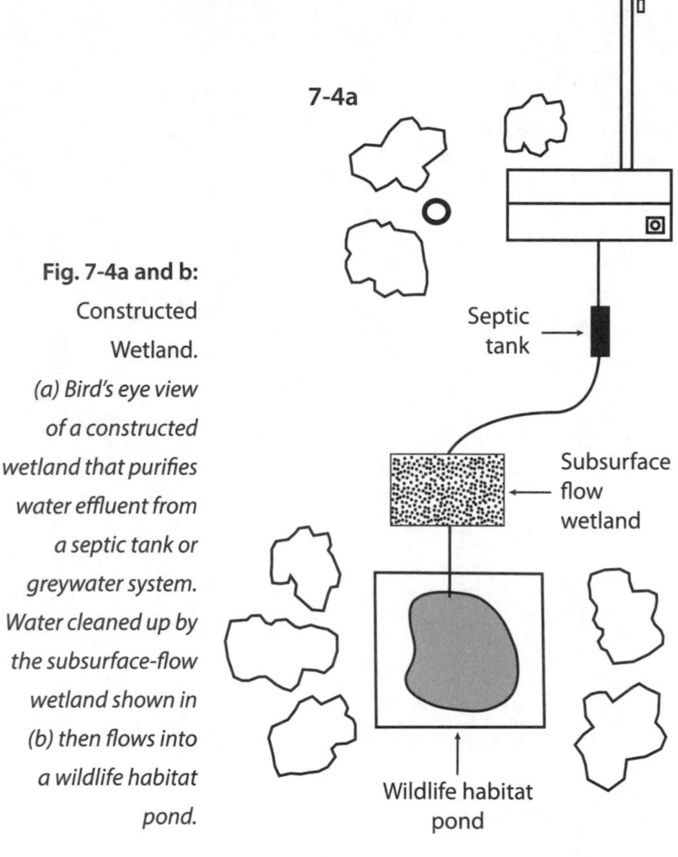

Fig. 7-4a and b: Constructed Wetland. *(a) Bird's eye view of a constructed wetland that purifies water effluent from a septic tank or greywater system. Water cleaned up by the subsurface-flow wetland shown in (b) then flows into a wildlife habitat pond.*

this application. I'll show you how you can incorporate a constructed wetland into your life, recycling nutrients from your home, farm, or business.

Residential Constructed Wetlands

When considering ways to capture nutrients from "waste" from your home, remember that constructed wetlands can be designed to capture water and nutrients from greywater and blackwater. They can be designed to service a single family residence or a group of homes — for example, a cohousing community, an ecovillage, or townhomes.

Many constructed wetlands are designed to be the primary recipient of household greywater. In these systems, greywater flows into the

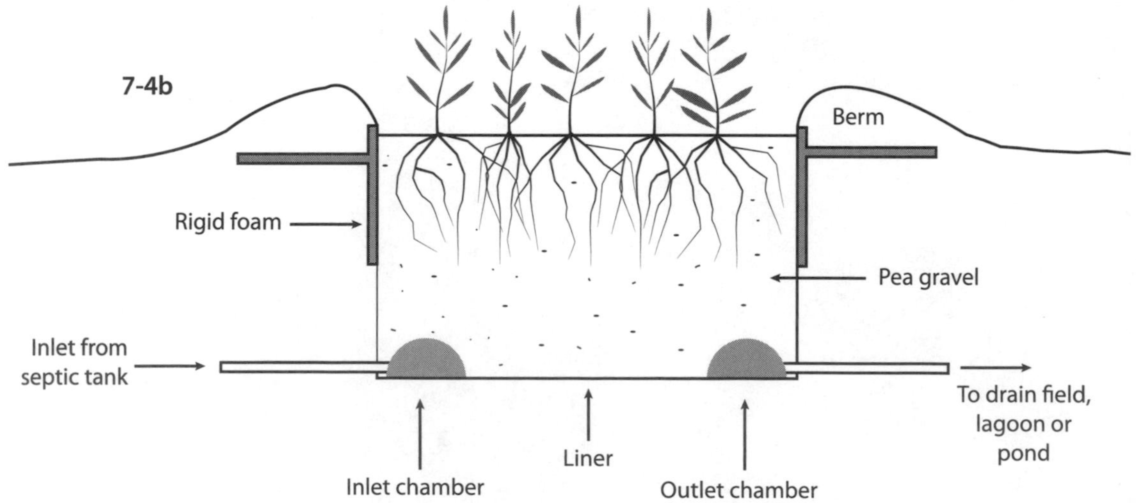

backyard wetland where it is purified. The effluent is then used to irrigate trees, shrubs, and a host of other plants. In some cases, the filtered greywater flows into small ponds that serve as wildlife habitat and/or recharge points for groundwater.

Constructed wetlands can also be installed to handle surplus greywater from backyard greywater irrigation systems. In these systems, discussed in Chapter 6, primary greywater nutrient and water capture occurs in mulch basins around plants. When greywater production exceeds demand — for example, when heavy rains reduce irrigation needs — greywater can be diverted to a small backyard constructed wetland rather than a septic tank or sewage treatment plant. If you are considering installing a greywater system and anticipating excess, give this option some serious thought.

Constructed wetlands can also be built to biologically process the effluent from residential septic tanks, as shown in Figures 7-4 to 7-6. In systems such as this, blackwater and greywater first flow into the septic tank. Septic tanks capture and hold solids, soap, and grease and a liquid known as "clarified water" from slightly below the scummy surface. Clarified water contains urine and some dissolved organic chemicals from feces. Its name is a far cry from reality — it is a brown, nutrient-rich liquid.

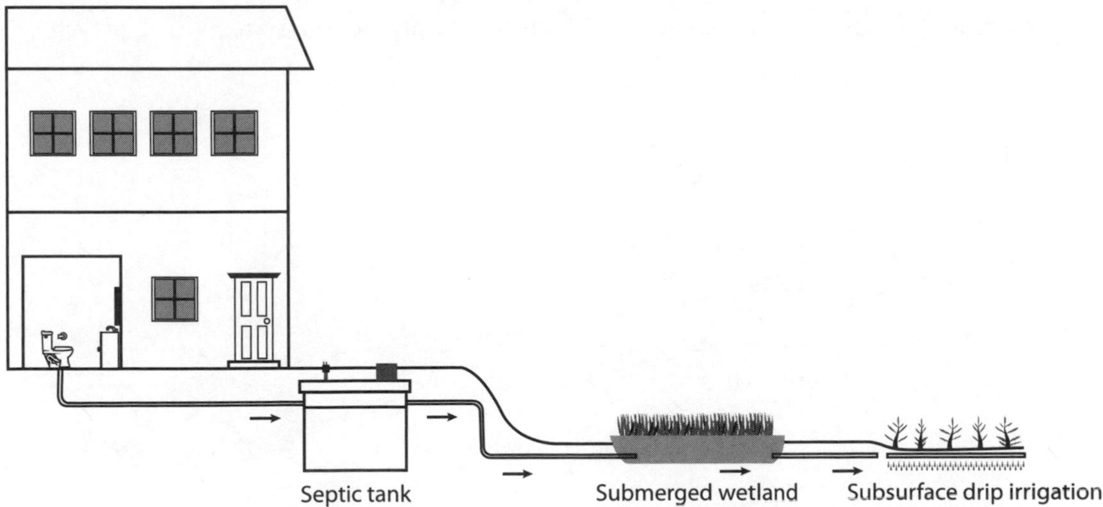

Fig. 7-5: Treating Septic Effluent. *The submerged wetland in the center of the drawing receives nutrient-rich effluent from a septic tank. It is purified in this constructed wetland and then fed into a subsurface irrigation system (shown here) or a wildlife habitat pond.*

Rather than being sent to a leach field, where it can seep into and pollute groundwater, septic tank effluent can be piped to a constructed wetland. The wetland could serve as wildlife habitat or may be designed to produce harvestable organic matter. The latter could be ground up and fed to animals or used as mulch in gardens.

Types of Residential Constructed Wetlands

Two types of constructed wetland are in use today: *surface-* and *subsurface-flow.* Both systems rely on natural biological processes to purify water, just like Mother Nature.

Surface-flow Systems

Shown in Figure 7-6, surface-flow systems are small-to-medium-sized ponds. Aquatic vegetation such as cattails, rushes, arrowhead, and pickerelweed are planted around the edges. The result is a pond ringed with aquatic vegetation. In the open water in the center of the pond, filamentous algae and duckweed grow. They can be periodically harvested and added to compost, dug directly into soil, or fed to livestock and poultry, provided they are free of toxic chemicals and pathogens.

As you may recall from Chapter 6, state and local health authorities assiduously guard against human or animal contact with greywater. As a result, they only permit systems that prevent or

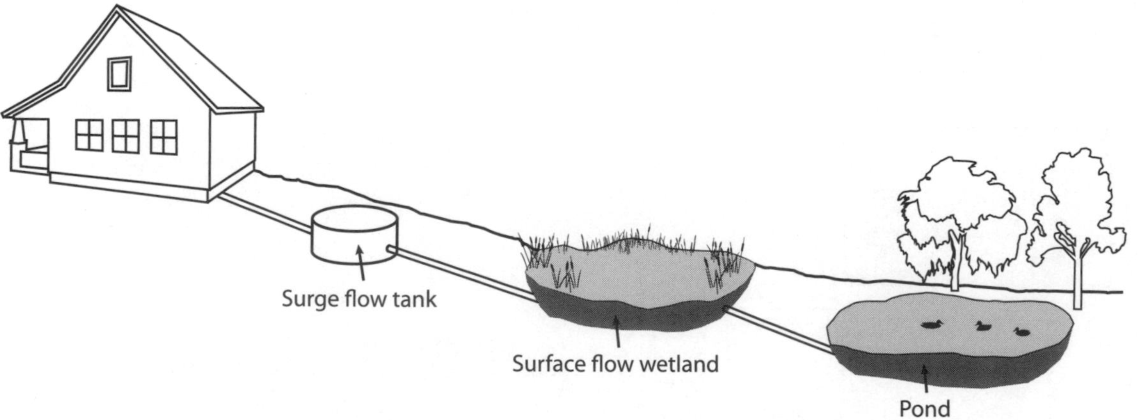

Fig. 7-6: Surface Flow Wetland. *Surface flow wetlands are popular in cities and towns, but generally not as widely used in residential applications.*

greatly reduce the chances of contact with greywater. Because of this, surface-flow constructed wetlands are rarely allowed for home "waste" water treatment. There's just too much risk that someone's child or pet will end up in the pond.

Most surface-flow constructed wetlands — and there are lots of them in use throughout the world — are used to purify "waste" water for sewage treatment plants of our cities and towns. These ponds are typically located on land that's a long way from human habitation and is often fenced off to prevent people and pets like water-loving Labrador retrievers from diving in to chase a ball.

If you want to build a surface-flow wetland, check with state officials. Many Midwestern states, such as Missouri, permit surface lagoons to process effluent from septic tanks. Lagoons provide a contained area that allows clarified water to evaporate. Solids settle to the bottom and eventually decompose.

Lagoons function year after year after year, faithfully wasting all those valuable nutrients in septic tank effluent. They are still permitted in many jurisdictions, however, because clay-rich soils render leach fields ineffective. That is, the soil greatly retards the percolation of effluent into the ground. If it can't percolate into the ground, effluent will often bubble to the surface of the ground above the leach field. Because of this, counties will allow lagoons to "process" septic tank effluent. Liquids evaporate from the pond.

If you live in a county or state that permits lagoons, you could easily make a case for the installation of a surface-flow wetland to treat greywater or leachate from your septic tank. Talk with your local and/or state authorities — whoever has jurisdiction over such matters. Start with your building department. They'll tell you who you need to speak with. Those jurisdictions that are willing to let you install a surface-flow wetland may be eager to gather data on their effectiveness. If you have a lagoon on your property, you might want to consider converting it to a surface-flow wetland. Local officials might be receptive if the wetland is well designed.

You will very likely need to hire a qualified engineer to draw up design and run calculations the AHJ (authority having jurisdiction) needs to ensure that the system will work under all conditions — that is, it won't overflow and pollute surface waters. A qualified engineer will design the system based on blackwater and greywater generation rates as well as precipitation and evaporation rates. You may also need to hire a professional to install the system, or work under a professional's supervision to qualify for a permit.

One warning: If mosquitos are a problem in your area, you may want to scrap your plans for surface-flow wetland. Open water provides excellent habitat for these pesky bugs, although we have never experienced this problem.

Lagoons can also be transformed into a subsurface-flow or submerged wetland, discussed shortly. Plants that grow in them can be periodically harvested and fed to livestock or poultry or used as mulch in gardens or to build soil in pastures and cropland.

Loving Lagoons

I first discovered lagoons when I purchased my farm in Missouri in 2008, which is now home to my educational center, The Evergreen Institute. Our lagoon received blackwater and greywater directly from the house. That is, there was no septic tank. I was amazed that the lagoon didn't smell at all.

When we rebuilt the house after a fire, the county required us to install a 1,000-gallon septic tank, even though we use composting toilets in the classroom building and the house. So only greywater would enter our lagoon.

If your home is serviced by a lagoon, you know that trees love lagoons — and grow prolifically along their perimeter. Unfortunately, building code officials view trees as a problem and may ask you to cut them down. They're very likely going to require you to do the same if you convert a lagoon to a wetland.

Their logic behind tree removal is that trees block sunlight, and, as a result, they reduce evaporation from lagoons. Unfortunately, officials don't seem to understand that trees suck up thousands of gallons of moisture through their root systems every year and release it into the atmosphere. In fact, trees very likely do a better job of removing water from lagoons than the sun! Moreover, trees can be periodically harvested for firewood or woodchips for animal bedding.

Subsurface-flow Systems

To prevent contact with greywater or the dark greywater effluent from septic tanks, most constructed wetlands are subsurface-flow wetlands, also referred to as submerged wetlands. If your county permits constructed wetlands for residential use, chances are they will only allow this type because they minimize the potential for human or animal contact with greywater and/or blackwater. They also eliminate the potential for turning your environmentally responsible nutrient recycling system into a breeding ground for mosquitos.

As shown in Figure 7-7, a submerged wetland consists of a porous medium, typically sand, gravel, pebbles, or small rocks, through which greywater or the effluent from septic tanks flows. The top of a rock-based subsurface-flow wetland usually consists of pea gravel, a medium into which one can plant seedlings. It could also consist of a layer of soil, but that comes with special considerations I'll discuss in the section on design.

While we are on the subject, many manuals and articles on subsurface-flow wetland systems recommend the use of gravel. Technically, the term gravel refers to small stones or a mixture of

Fig. 7-7: Subsurface-flow Wetland. *This drawing shows the size of rocks used to form a biological filter for greywater. One of the keys to success is ensuring gradual, but continuous flow and preventing interstices from becoming clogged by sediment or organic matter.*

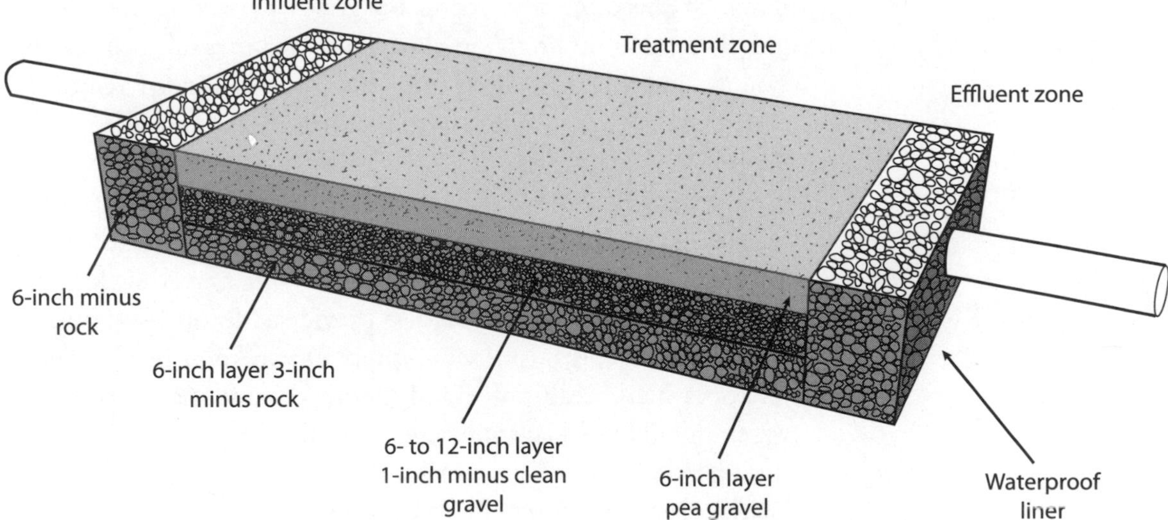

small stones and pebbles with sand. Gravel may also contain silt and clay. Unfortunately, silt and clay fill the spaces between small rocks in gravel in a submerged wetland. This may retard water movement. Remember: when designing a submerged wetland, your main goal is to create a porous substrate — that is, a material with lots of spaces that allow water to flow through the medium. Porous materials also create lots of surface area for microorganisms to set up shop. Microbes form extensive colonies on the surfaces of the rocks, where they dutifully digest organic matter.

Be very specific when you order crushed rock. For optimum longevity and performance, only use *clean or washed* crushed rock — that's rock with little, or no, sand and/or *fines* (fines are particles generated when rock is crushed). Three-inch minus clean rock work wells. The designation "clean" or "washed 3-inch minus" rock means that the rock in the mix will be in the 2–3 inch range — and should contain very little sand or silt or clay fines. Talk it over with your supplier so he or she is fully aware of what you want and be sure to repeat the word clean or washed several times. I've had haulers deliver unwashed rock when I had ordered washed rock. It's a good idea to view the product in person prior to trucking it to your place.

Whatever you do, don't take a front-end loader or skid steer down to the creek and scoop up gravel to save money. River gravel can clog up pretty quickly when used in a subsurface-flow system, and you'll be faced with the onerous task of removing the gravel and replacing it with the right stuff. (Call in Mike Rowe!)

In a subsurface flow wetland aquatic vegetation is planted in the pea gravel or sand perched atop the gravel filter (Figure 7-7). Vegetation sends its roots into the underlying gravel, sucking up moisture and nutrients. Roots also provide additional surface area on which microorganisms set up shop. As you know by now, these microbes help break down and absorb organic materials in the water and liberate plant nutrients from the "waste" water flowing through the gravel. The more surface area that is covered with bacteria and other useful microbes, the better.

Plant roots also deliver oxygen to the root zone. Oxygen is required by cells of the roots. And, it also helps support aerobic bacteria that decompose and recycle nutrients in the water.

Subsurface-flow systems are designed so that water will remain about 3 inches (7.5 cm) below the surface at all times. This reduces the potential for contact with bacteria- and virus-laden water. It also helps to prevent odors.

Be sure to build the banks of your wetland high enough to accommodate a rapid influx of water from torrential downpours. Also be certain that your design doesn't allow surface runoff from neighboring areas to flow into the wetland. Surface runoff can add a lot of unwanted water and pollutants. Excess water could cause the wetland to overflow its banks, sending your greywater onto the surface of the ground or into nearby streams, ponds, or lakes.

Types of Subsurface-flow Wetlands

Subsurface-flow constructed wetlands can be of two types: *lateral flow* or *vertical flow.* Figure 7-7 shows a lateral-flow design. In this system, water enters the upper end of the wetland, then flows through the porous substrate. It is physically and biologically filtered and purified as it flows to the lower portion of the wetland. The residence time, that is, the amount of time the water spends in the system and is "biologically scrubbed," depends on the design. Most designs call for two to three days residence time. I'd suggest a five-day residence time, as it permits more time for microbes to do their thing.

By the time the water reaches the lower end of the slightly sloping subsurface-flow wetland, it should be pretty clean. If so, it can be safely released into a small polishing lagoon, a specially constructed wildlife pond, or an irrigation system like those described in the chapter on greywater. It could also be disbursed by an underground leach field, although this drastically reduces the utility of the water you've just scrubbed.

The choice of the next stop in the water's course after it leaves the wetland depends on your preference and, lest we forget, on

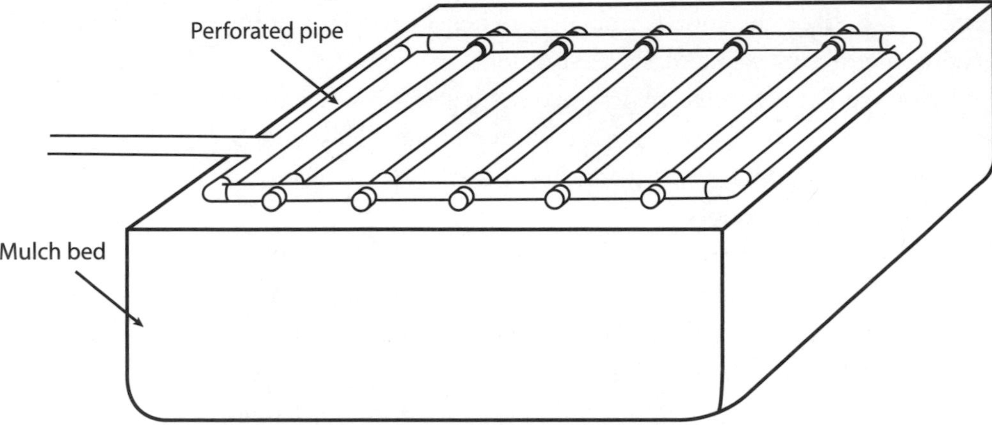

Fig. 7-8: Mulch Basin. *This vertical-flow wetland acts like a huge sponge that absorbs greywater.*

state and local regulations. In Nebraska, for instance, state law prohibits discharge directly onto land or natural bodies of water unless a permit is obtained. In this state, you'll need to send the purified greywater to a lagoon, wildlife pond, or leach field.

States may have rules that govern the infiltration rate of wildlife habitat ponds — that is, how fast water can seep out of the pond and into underlying soils and groundwater. A professional engineer or a knowledgeable professional installer can help you figure this out. You may also be required to fence off the pond.

The second type of subsurface-flow wetland is the vertical flow. As shown in Figure 7-8, in this system pipes release the greywater onto the surface of mulch or some other porous organic material housed in a very large mulch basin. The water soaks into the material and then into ground. Plants growing in the system can then soak up the nutrients.

Most systems in use today are lateral flow.

Principles of Design and Construction of Subsurface-flow Wetlands

With these basics in mind, let's take a closer look at constructed wetland design. We'll focus on subsurface-flow systems because they are more likely to be approved in your locale. Once you have read and studied the material in this chapter, you may want to go online for more details. There you'll find a lot of information on subsurface-flow systems through the US Environmental Protection Agency, permaculture organizations, the University of Nebraska, Colorado State University, University of Nevada Extension Service, University of Minnesota, wetland construction businesses, and so on. Also be sure to contact your local building department. They'll probably send you to the state water office. They may refer you to the state extension office.

Rather than provide detailed descriptions of system designs, which will vary from one authority to the next, I'll review key design principles. This section will provide a fairly detailed description of features you'll need to incorporate or at least consider when designing and building a safe subsurface-flow wetland.

Gravity Flow

As with the greywater systems discussed in Chapter 6, design your wetland system for gravity flow. Let gravity move water throughout the system — from the house to the septic tank (if any) to the wetland to the irrigation field or wildlife habitat pond. Pumps can be costly, require periodic cleaning and maintenance, may require periodic replacement, and, of course, use energy that costs money and results in environmental pollution.

To permit gravity to work, be sure to provide adequate fall (drop) in pipes — usually a ¼ inch per foot, or 2%, drop is sufficient. Be sure the wetland itself slopes slightly as well, but not too much. Generally, slopes up to 1% — 1 foot in 100 feet (0.33 meters in 30 meters) are adequate. Remember: you want water to flow through the system, but slowly. This permits adequate time for natural filtration, microbial decomposition and conversion, and uptake by the roots of wetland plants. As just noted, two to three days are generally recommended, although slightly longer retention times — up to five days — may be required if you are discharging water into a wildlife habitat pond and processing greywater all year long, including the winter.

Take your time when digging the wetland and shaping the bottom, and be sure you achieve a nice, gradual slope. A 4-foot level and a long, straight 2 x 4 will help ensure you are correctly sloping the bed. Use a garden rake to create an even, rock-free surface. Be sure to remove rocks, especially sharp-edged stones, if you are going to install a synthetic liner in your wetland.

Sediment

Design the system to prevent sediment from entering. Sediment may be released by erosion of the berm or from neighboring areas. To prevent sediment from "the banks of your own lagoon" (I had to include that reference from The Beatles' song) from entering the wetland and clogging up the spaces between rocks, be sure to vegetate the banks as soon as possible. Build your wetland in the spring after the rains have passed, but not in the dead of summer when it's difficult to get plants to grow. To help establish a thick, water-absorbing soil-retaining vegetative cover, consider seeding the banks with fast-growing plants such as wheat, rye, or oats. They grow quickly and provide protection from erosion right away. If you use grass, be sure the soil you are sowing seeds into is rich in organic matter. This will increase the likelihood that the vegetation will take hold quickly. Grow native grasses, too, as they're adapted to the amount of precipitation and soils in your area. Cover newly planted seeds with layer of chopped straw. Chopped straw or some other type of mulch often helps plants get a start, as it helps retain soil moisture that is so essential to seedlings when they first set sail in this world. Re-sow bare spots — and get to them quickly. Water as required.

Although it is best to design and build a wetland that does not drain other areas to reduce sedimentation and potential flooding, you can design one to treat water from surrounding areas. However, to ensure longevity, be sure to minimize erosion from neighboring areas. Seed bare areas in the watershed of your catchment area, for instance. Create a sediment deposition basin or two upstream from the wetland as well, if necessary.

When designing and building a wetland, be especially sure not to let water flow into the wetland from nearby parking lots or driveways — as they may contain oil from cars and trucks. Also, don't drain areas that are sprayed with herbicides or insecticides.

Liner

To prevent water from leaking out of your wetland, line the bottom with a waterproof material. Typical liners include clay (such as bentonite), concrete, rubber, or plastic. Native clay may be sufficient to hold water. If not, adding a thick layer of powdered bentonite may suffice. Bentonite is a highly expansive clay. It swells to about 19 times its volume when wet, creating a watertight seal at the bottom of ponds (if a thick enough layer is applied). You can purchase bentonite in bags at local feed stores and in bulk. Call local pond builders for bulk sources.

Synthetic liners can be made from a variety of materials, such as high-density polyethylene, PVC, polypropylene, or EPDM (a rubber material). Be sure to order a thick liner — in the range of 20 to 30 mils. Forty-five mil UV-resistant butyl rubber will work. (Butyl rubber is a synthetic rubber.) I have used UV-resistant EPDM, another synthetic form of rubber, to make shallow duck ponds with great success. (EPDM stands for ethylene propylene diene monomer.) I purchase EPDM online from Webb's Water Garden, but there are other suppliers as well. (Search under pond supplies.)

EPDM is thick, durable, and easy to work with. It is costly, however, and will set you back a bit if you are building a large wetland. UV resistance is important if parts of the liner around the edge of the wetland are exposed to sunlight.

When installing a synthetic liner, also consider applying a protective layer of geotextile material below it to prevent puncture. This material can be purchased at pond supply companies. A 2-to-3-inch layer of sand (5 to 7.5 cm) below the liner may be sufficient to protect the liner from being punctured by sharp rocks. Be sure to tread lightly on the liner during installation to prevent damage.

If the pipe going into or exiting the wetland penetrates the liner at any point, be sure to install a fitting called a bulkhead

Fig. 7-9: Bulkhead. *This handy device is used to penetrate tanks and pond linings, creating inlets and outlets.*

(Figure 7-9). First described in Chapter 4, bulkheads are watertight fittings meant for penetrating walls of tanks or synthetic liners. They can be attached to plastic pipe and can be purchased online and at farm supply stores.

Berms

Be sure to build a watertight berm around the wetland to prevent overflow. A 2-foot-high (about ⅔-meter) berm may be sufficient, but by all means design with local rainfall in mind and consult local regulations or guidelines from state university extension agencies (or other sources) for recommendations on berm heights in your area. Also, be sure the berm has a solid clay core to prevent leakage, and don't let trees such as willows grow on the berm. Remove them as fast as they sprout. However, be sure to avoid herbicides such as Roundup.

On our farm, I've controlled annoying and prolific willows on pond banks by cutting them each year — a thankless and discouraging task, as they always come back. I've dug some up, but that can be a chore. The best control measure I've found is to burn the willows. I first trim the small willows down to the trunk with loppers or pruning shears. I then build a teepee fire around the remaining trunk and root crown, using waste wood from my workshop. I place some paper inside the teepee and douse the wood and paper with biodiesel, then light it. For best success, I let the fires burn for an hour or two. That kills these resilient plants. The first time, I burned nearly 30 willows growing on the dam of my pond. I found that about a third of them sprouted back to life because I missed a branch or a root. A second burning killed all but one. A third burning completed the job.

For safety, burn after rainstorms while the surrounding grasses are still damp. Or, spray water on the grasses around the tree *before* you light the fire. Be sure to have a water hose or buckets of water nearby to douse fires that may pop up. It goes without saying, that you should never leave while a fire is burning or burn on windy days.

Positioning Your Wetland

Another important rule is not to build a wetland in a flood plain or arroyo. Flood plains are the lowlands that border rivers, streams, and creeks. They periodically flood, often during heavy rains or when heavy winter snows melt. Arroyos are dry gulches in the West that fill with water during infrequent rainstorms.

Don't, under any circumstance, build a wetland in a floodplain or natural drainage like an arroyo. At some point, it will flood and could compromise the integrity of the berm, spilling greywater into a natural waterway.

Sizing Your Wetland

Size the wetland properly. According to an online publication from Perdue University (Indiana), wetland cells should be sized for five-day retention. To meet this goal, the system should hold one gallon per square foot (about 4 liters per 0.09 square meters). Remember, though, requirements vary, depending on evaporation rates and rainfall. Consider this a guideline. Seek site-specific advice.

Most subsurface-flow wetlands are 18 to 24 inches (46 to 60 cm) deep. The length-to-width ratio is typically 2:1 to 3:1. According to Perdue University, a two-bedroom home in Indiana will require 300 square feet (a little less than 28 square meters) of wetland. If the length-to-width ratio is 2:1, the system could be 25 feet long and 12 feet wide (7.6 meter x 3.6 meters). A 30-foot long by 10-foot wide (9 meter by 3 meters) wetland would work well, too. That's a 3:1 ratio.

A constructed wetland for a three-bedroom home should be 450 square feet (42 square meters), or 30 feet long and 15 feet wide (9 meters x 4.6 meters) to maintain a 2:1 ratio. This estimate is based on the use of up to 750 gallons (2,839 liters) of water per day in Indiana. Building based on this estimate will also ensure that the system will continue to operate during winter months — if you build it right. Yes, it's true that constructed wetlands will continue to purify water during winter months. A study conducted in Hungary showed that reeds (cattails) in constructed wetlands continued to process organic substances in greywater throughout the winter with very little loss of efficiency.

But don't forget: when designing a system, its size should be based on conditions specific to your area and the amount of greywater you produce. Household water use doesn't depend on the number of bedrooms, but rather the number of people and, even more important, per capita water consumption — how much water each person uses. Remember, bedrooms don't use water, people do.

Avoid Clogging

For longevity, be sure to design and build the subsurface flow bed to ensure continuous water flow. Translated: design the system to prevent clogging. The medium through which greywater and dark greywater flow needs to be carefully selected.

When water first enters, it has the highest content of crud: organic matter, fibers from clothes, hair, skin cells, and so on. This region of the wetland therefore needs to have the largest channels. This prevents materials from clogging the spaces between rocks. Perdue University's publication on greywater recommends

1.5-to-3-inch (4 cm to 8 cm) rock at the inlet and outlet of the system. Six inches (15 cm) of river rock along the bottom of the wetland also facilitates the flow of water from inlet to outlet.

Between the influent and effluent sections of the wetland, say the researchers from Perdue, there should be 0.5 to 1.0 inch (1.25 cm to 2.5 cm) clean crushed rock, free of silt and sediment. It should run to within 6 inches (15 cm) of the surface.

The last 6 inches (15 cm) should consist of pea gravel — rock that's ⅜ to ½ inches (1 to 1.25 cm) in diameter. Pea gravel can be purchased by the bag at home improvement centers and garden centers. I have found that bagged pea gravel is pretty dirty, however, so I recommend that you wash it with a hose in a wheelbarrow prior to layering it into a constructed wetland. It's a very good idea to wash down clean gravel as well — just to be sure it's clean. You can build a sieve out of ¼-inch hardware cloth and wash it down that way or simply pour the contents of several bags into a wheelbarrow and hose it down. Add water, then stir it around with your hands, and let the dirty water flow over the rim of the wheelbarrow. (This is a job that goes a lot faster with two people.)

Washing gravel will slow the project down, but it could mean the difference between a lifetime of service and a lifetime of costly maintenance and headaches. If your system clogs, you'll need to pull up the plants, remove the substrate, extract roots, and wash it or dump it somewhere safe and start over.

Frankly, I'd be very leery about using sand. I have no experience with this substrate in constructed wetlands, but my guess is that it would clog pretty quickly over time. While you are at it, it is a good idea to locate constructed wetlands away from deciduous trees, if at all possible, so fall leaves aren't deposited in it.

Another way to minimize clogging — and reduce maintenance — at the inlet is to create two or more points of discharge into the wetland. Less crud is absorbed by the rock filter at each location. Bacteria should be able to process the materials and keep the interstices — the gaps between rocks — from clogging.

In larger systems, you might consider creating a notched weir — a shallow trough that runs the width of the wetland. Notches in

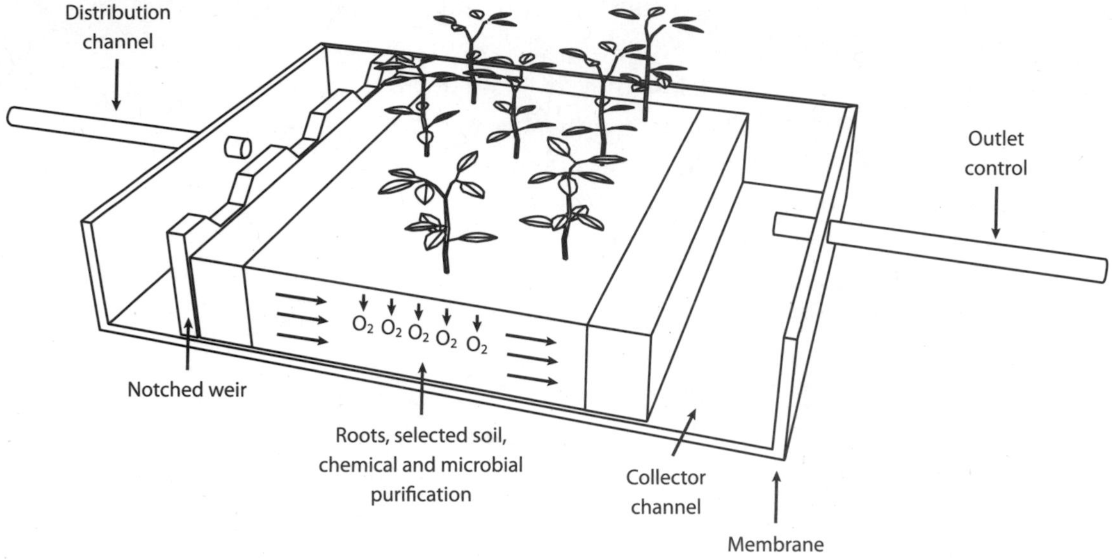

Fig. 7-10: Notched Weir. *This concrete structure helps disperse the flow of greywater into a submerged wetland, greatly reducing the chances of organic matter from blocking the gaps between rocks.*

the weir (the dam on the inlet side of the wetland) allow the water to trickle into the wetland at several points (Figure 7-10).

Insulation

Submerged wetlands can function throughout winter months in warm as well as cold climates — if properly sized and designed. In climates that experience freezing weather, design the system to retain heat. This can be achieved by insulating the banks or berms of your wetland. Two to four inches (5 to 10 cm) of rigid foam insulation that's rated for burial will help prevent heat from escaping from the wetland system and prevent it from freezing. I'd recommend extruded polystyrene (pinkboard or blueboard), which is rated for burial, not polyiso or expanded polystyrene (beadboard), which are not.

In an online article entitled "Constructed Wetlands" written by researchers at Colorado State University, the authors recommend raising water levels in both types of constructed wetland by about 18 inches (46 cm) in the late fall or early winter until a sheet of ice forms on the surface. When that occurs, drop the water level. This creates an air pocket between the sheet of ice and the top of the water level. This, they contend, will help insulate the biologically active zone of the wetland. How you would do this in a submerged wetland without bringing potentially contaminated water to the surface escapes me and is a reminder to read what's on the Internet with a critical eye.

Aeration

Consider installing some form of aeration. Some experts on residential constructed wetlands recommend aeration. (See the online article published by the University of Nevada Cooperative Extension entitled "Constructed Wetlands for Water Filtration.") Aeration, say these individuals, increases oxygen in the water that helps promote aerobic decomposition. It also helps supply plant roots with much-needed oxygen. Because roots absorb nutrients and water and provide abundant surface area for microbial colonization, an aerated system may perform better.

Aeration systems require electricity, however. If you want to aerate, consider installing a solar-powered aerator. I installed one on our pond in conjunction with John Redd from Water Solutions, a St. Louis-based company that now sells these systems online. Although aeration may help, aeration stones could become clogged with organic sediment and gunk, necessitating periodic cleaning. Make sure they are easy to access, just in case.

Planting and Maintaining a Constructed Wetland

When designing and building a wetland, be sure to do so in ways that minimize clogging and maintenance. Some of the design principles just discussed serve that goal. Choice of plants and annual plant maintenance will also help reduce clogging and the need for major maintenance down the road.

When planting a constructed wetland, you'll draw from three groups of plants (called genera): *Typha, Scirpus*, and *Phragmites.*

One of the most familiar wetland plants belongs to a group (genus) called *Typha.* This genus includes about 30 species. Its members can grow to 6 feet (1.8 meters) tall — or more. If you live in North America, you may be familiar with one of the most prominent members of this genera. We call them cattails or reeds.

If you are from Great Britain, you may know them as bulrushes or reedmace. In Australia, they're referred to as bulrushes and cumbungi.

When vegetating a wetland, select native species — plants that are indigenous and hence well adapted — to your region. Plant a variety of species. Diversity in ecosystems typically leads to greater stability. If one species fails to thrive, no sweat, there are plenty of others. Plant diversity also results in greater animal and insect diversity. The more plant and animal (bird and insect) diversity, the better.

For advice on suitable wetland plants, contact your state's Division of Wildlife or Department of Natural Resources. Ask to talk to a pond or wetland specialist. He or she may be able to give you sound advice on aquatic plants that will do well in your area. You may also want to contact the Cooperative Extension office of state universities. They may have a list of suitable aquatic plants. It might be worth emailing the person rather than talking in person so you get a written response. Experts can spout off species with confusing Latin names faster than your brain can process the information! They are often in a hurry, too. State wildlife and natural resource conservation agencies may also have manuals on pond maintenance that list suitable wetland plant species.

Local nurseries may sell a limited number of aquatic plants for backyard ponds. For example, they may sell species like arrowhead, pickerel weed, and cattails — although they tend to be pricey.

You can also transplant aquatic plants from neighbors' or friends' ponds. (I've transplanted many in my time!) Aquatic plants are generally very hardy and transplant well. Once established, many spread via their roots.

You may also find aquatic plants online through Craigslist. I purchased dozens of pickerel weed to re-vegetate our farm pond — at a very reasonable price. (It had been devastated by a beaver.)

You may also be able to harvest aquatic plants from public waterways and lakes. But be sure to check regulations *before* you go plant napping! You don't want to end up in jail for robbing cattails or pickerel weed. (The murderer with whom you share your jail cell won't be too impressed.)

Also, be sure to select carefully, especially if you are building a surface-flow wetland. You don't want to introduce species that will take over. Consult pond management manuals on weedy species that have a tendency to take over and choke a pond. Avoid them at all costs!

Whatever you do, do not import plants from one part of the country to another. You may inadvertently introduce a species that becomes a pest in its new habitat.

University of Nevada Area Extension Specialist M. L. Robinson and coauthors of "Constructed Wetlands for Water Filtration" (available online) make a case for growing nonaquatic plants in constructed wetlands. Since you are essentially creating a hydroponics gravel-filled grow bed, they argue that "almost any plants, even nonaquatic, including drought-tolerant plants such as rosemary, sunflowers, and others, will grow" in these conditions. Plants that bear flowers add beauty and provide food for birds and other wildlife. Select wisely, however, as these plants need to be species whose roots can be immersed in water 24 hours a day. Not all species like their feet wet all the time. Whatever you do, don't plant trees or shrubs. Their roots can take over the system and may puncture the liner, causing it to leak.

Also be sure to inspect the plants in your wetland — especially early in its life — for signs of stress, insect infestation, or disease. It may take you a while to learn the natural cycles of plants and what's healthy and what's not, but over time, you should be able to determine when your wetland crop is doing well and when it's not.

Keep a close eye on your plants. But don't expect garden-perfect plants at all times. Some leaves will turn yellow and die. That may be natural. Some insect damage — like chewed leaves — is inevitable as well. That's nature. However, if large numbers of plants are withering, turning yellow and dying, you have a problem. If you have questions about plant health, contact a state official who is an expert on wetland plants. He or she might be willing to visit your site to inspect your plans and make suggestions. Or, they might be willing to examine photos or specimens you send or take to his or her office. (That's more likely.)

One bit of advice when planting: buy older (larger) plants. One-to-two-year-old plants are more likely to survive transplanting, and they'll help you establish your wetland more quickly than seedlings.

A surface-flow wetland won't operate optimally until the vegetation and the layer of organic matter that forms above the roots (known as litter) are fully developed. This could take one to three years, depending on a number of factors — for example, when plants are set out, the age of the plants, and how closely they are planted. If you start with older plants early in the spring and plant them fairly closely, you could have a pretty decent wetland by the end of the first summer. Certainly by the end of the second summer your system should be operating at full tilt.

Drought can make it difficult to start and maintain a constructed wetland. Water shortages are a major plant stressor. During long, dry spells, homeowners may need to irrigate their constructed wetlands using tap water or water collected off rooftops. Rainwater will reduce one's use of well water or city water — and save money and energy. Store rainwater in a buried tank or a dark-colored above-ground tank to prevent algae from growing in the tank.

Be sure to check the water level in your wetland every couple of weeks, especially during dry spells. Do so by scraping away some of the pea gravel on top aside to locate the water line. Better yet, install a vertical perforated plastic pipe in the gravel bed. Run it to the bottom of the bed. Perforations along its length to allow water in. Place a small cap on the pipe that can be removed to check water level.

Watch for invasive species — natural or foreign — that become established in your wetland. Also be sure to remove grasses, shrubs, and trees that pop up on their own. Pull them up carefully, root and all! To remove roots, pull slowly and steadily near the base of the plant. This will allow you to extract as much root biomass as possible. Never cut off undesirable species, as they'll probably just grow back from the roots you left behind, and they'll often be even hardier!

If you live in a temperate climate where plants go dormant in the winter, you may want to harvest some of the plant biomass each year. This prevents plant matter from building up and creating a thick layer of organic material on top of your subsurface-flow wetland. Decomposed organic matter could ooze into the filter medium below it, causing it to clog and reducing the lifespan of your system.

Trim back vegetation with sheers, hedge trimmers, or some similar tool. A weed whacker may prove helpful; however, be sure that you remove all the cuttings. You may need to rake them up.

It may also be necessary to thin plants from time to time as well, if your system becomes overrun with vegetation. Be sure to pull plants up root and all. Plants harvested from a constructed wetland can be sold to other like-minded individuals. If your system is successful, you could supplement your income by providing aquatic plants for others!

Economics of Constructed Wetlands

A professionally installed septic tank and constructed wetland will cost more than a traditional septic tank and leach field — or a septic tank and a lagoon. Although the price varies, a septic tank, including installation of a leach field, can easily cost $3,000 or more. Add another $1,000 to $4,000 for a wetland, if you hire a professional. Bottom line: be prepared to pay a bit more to choose a more environmentally responsible option.

Higher initial costs can be offset by many noneconomic factors — you know those things we care about but on which we can't put a price tag. One of them is the satisfaction of reusing nutrients and water from your home. That's priceless. The beautiful carbon-absorbing vegetation that graces your wetland and helps curb climate change are two other nonmonetary benefits. So is the wildlife habitat you create. Graceful hummingbirds and colorful butterflies that flutter among brightly colored flowers in your backyard habitat add immense value. Irrigation of trees and shrubs from the effluent of a constructed wetland are additional factors that make this option profitable even though it doesn't pencil out that way.

Although you may not be able to quantify all the benefits of a constructed wetland, the art of conscious, meaningful existence requires the ability to understand the value of actions that don't come with a price tag. On a personal note, I was sitting on my patio one day when a red-winged blackbird flew directly at me from my pond, showing off those bold red shoulder patches. I have never seen the bird from that angle and had, as a result, never appreciated how magnificent those red wing patches truly were. We don't get to see them straight on very often.

Pros and Cons of Constructed Wetlands

Constructed wetlands offer many benefits to homeowners and to the environment, the life-support system of the planet. They are, for example, far more ecologically sound than septic tanks and sewage treatment plants.

Moreover, if designed correctly, wetlands systems produce no pollution. They rely entirely on gravity and clean, free solar energy, which powers the plants. No fossil fuels are required to pump or treat wastewater. No harsh chemicals are required, either. Moreover, constructed wetlands allow you to recharge the groundwater around your home. This will help supply the trees in the vicinity with the water they need to survive and reproduce.

You can use the output of a constructed wetland to irrigate plants or to simply create a delightful, species-rich wildlife habitat pond that serves wildlife and becomes a source of joy and entertainment to you. Finally, wetlands also allow us to grow plants that can provide food for livestock and organic matter to enrich garden soils.

On the negative side, you'll very likely need to monitor a subsurface-flow wetland fairly carefully. That is, you'll need to keep an eye on plants, be mindful of weather (mostly temperature and rainfall), and check water levels in the gravel bed every other week or so during dry spells. During such times, plants may become inordinately stressed and wither and die. To prevent this, add water *before* signs of stress appear. This will require vigilance on your part.

To me, the biggest downside of constructed wetlands is that they don't allow us to capture as many nutrients as composting toilets and the greywater systems discussed in previous chapters. Sure, wetlands will produce an abundance of aquatic vegetation, but if the blackwater and greywater first go to a septic tank, a composting toilet and a greywater system will provide a lot more nutrients — and more direct benefit.

You may also find a constructed wetland more challenging to permit. You may need to hire a professional engineer to design the system and a professional installer to put the system in to qualify for a permit.

Despite the downsides, a constructed wetland may suit your needs perfectly and provide you with the satisfaction of becoming a better steward of the Earth's resources. You can always add a composting toilet at a later date.

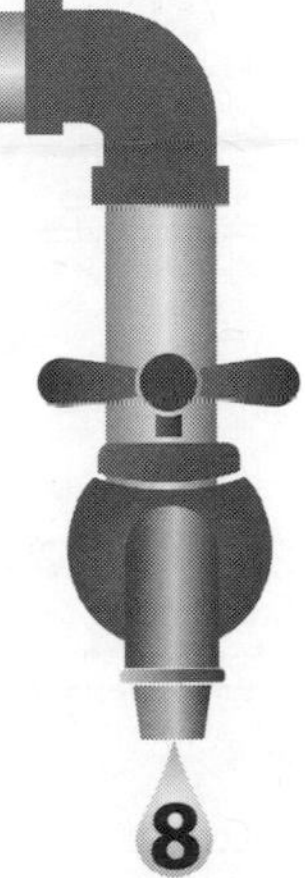

Greywater Planters

One of the latest crazes in food self-sufficiency is aquaponics. As you may know, aquaponics systems are designed to permit individuals to grow vegetables — indoors or outdoors — in a watery medium enriched by nutrients supplied by fish. In this system, one of several varieties of fish, such as tilapia, are raised for food. The fish are housed in a special tank near grow beds. Water from the tank contains organic and inorganic "wastes." Water from the fish tank is cycled through the growing beds of the aquaponics system. The nutrients are absorbed and used by plants to produce a wide range of edible vegetables. The plants therefore serve two key functions: (1) they remove nutrients and purify the water and (2) they produce edible food — much like natural aquatic ecosystems.

Aquaponics is a great way to provide fresh vegetables such as lettuce, Swiss chard, basil, and spinach throughout the year. For meat eaters, it also provides some delicious fish protein. What's amazing about aquaponics systems is that they can be set up in a tiny apartment in the heart of a major city, although the plants will very likely require artificial lighting.

As shown in Figure 8-1, in aquaponics systems, vegetables can be grown in beds of pea gravel or special clay pellets. They can also be grown in specially designed pots (net pots) suspended in the nutrient-rich aquarium effluent. The pots are placed in holes in rigid foam that floats on the water's surface.

On our 60-acre farm in east-central Missouri, we raise vegetables in several gardens as well as a medium-sized greenhouse.

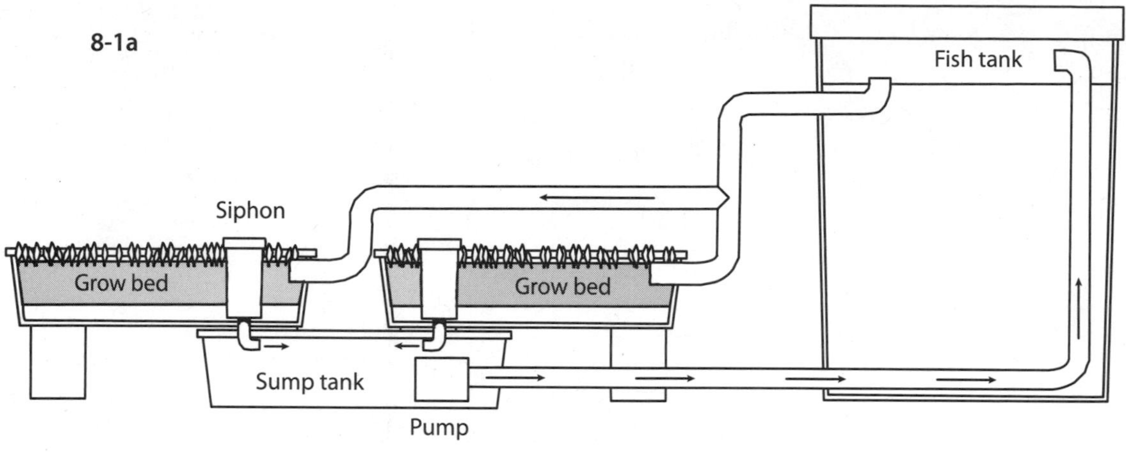

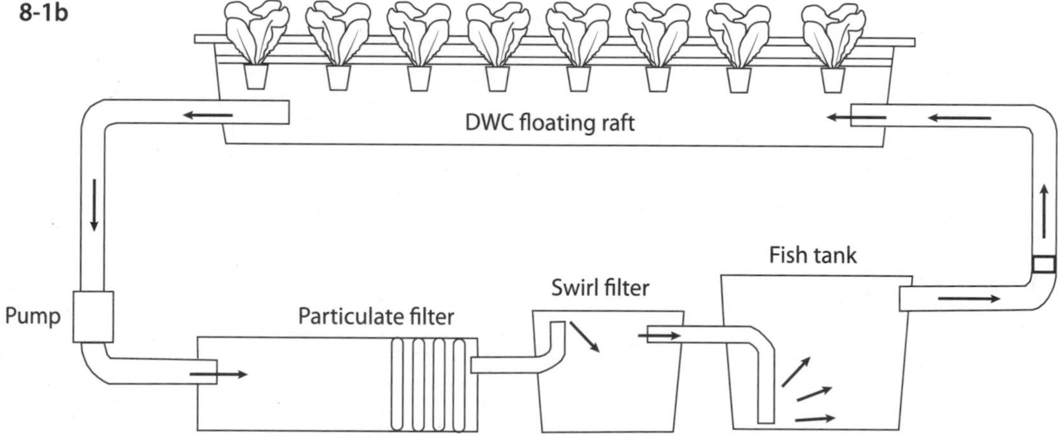

Fig. 8-1a and b: Aquaponics Grow Beds. *Aquaponists grow a wide variety of vegetables on (a) pea gravel or other similar media or (b) in pots suspended from rigid foam insulation rafts. Nutrients are derived from a nearby fish tank.*

Having battled weeds for years, we recently decided to transition to aquaponics. Our system, however, uses "waste" water from a pool our ducks frolic (and poop) in. (Ducks can pollute water faster than any other species on planet Earth!)

This system, which we refer to as "duckaponics," provides vegetables for us but also provides food for the ducks and chickens we raise on our farm. We are still perfecting the system, but have already successfully grown lettuce, spinach, chard, and turnip greens. What is more, we also grow large quantities of filamentous algae and duckweed. Ducks are especially fond of these foods.

To sum up, in an aquaponics system, "waste" water cycles between the growing beds and the fish tank or, in our case, the duck pond. Nutrients are removed by bacteria and other microorganisms living in grow beds. These organisms make the nutrients removed from the "waste" water available for the plants.

With a little creative thinking and ingenuity, you can set up a similar system using greywater. This system, if designed correctly, will allow you to grow nutritious food year round in planters nourished by greywater from your washing machine, sinks, and perhaps even your shower — as long as you use the right kinds of soaps and other cleaning agents.

Before you race out to buy a book on aquaponics, let me point out that there is a major difference between an indoor greywater planter and an aquaponics system. The key difference in an indoor greywater system is that plants are typically grown in topsoil — not in rigid foam rafts floating on water or in pea gravel. Topsoil is placed over a bed of rocks through which the greywater flows, much like a submerged wetland.

Another key difference is that aquaponics systems tend to be fairly closed systems — like ecosystems. Water is cycled continuously through the system. It's not *entirely* closed, however, because nutrients are introduced to the system in the form of fish food.

In greywater planters, water is not typically cycled over and over through the system. It is purified and then used for something else. That said, water could be purified enough to be reused — for showers, baths, cooking, dishwashing, or washing clothes — but it typically isn't. In most cases, purified greywater is used to water indoor or outdoor plants. It is also often cleaned up enough to be used to flush toilets. As a result, in greywater systems, water typically flows in one direction through the system. This is referred to as an *open system,* and it will help you achieve

Indoor Greywater Planters

Indoor greywater planters are biological filters, like aquaponics systems. And like aquaponics systems, they, too, can produce delicious vegetables year round, even in a tiny apartment in New York City or a suburban home in North Carolina!

self-sufficiency and live a life that is more in tune with Nature's laws of survival.

In indoor greywater planters, the roots of vegetables, miniature fruit trees, and other plants grow into the soil but also dip into the underlying greywater-saturated rock bed. Bacteria and other microorganisms growing on the surfaces of the rocks and the roots of the plants digest organic matter in the greywater, liberating nutrients that nourish plants — as in a submerged wetland, the topic of the previous chapter.

In this chapter, you'll see how indoor and outdoor greywater planters are built, how they function, and what's required to achieve success. Let's begin with some options.

Indoor Greywater Planters Options

Indoor greywater planters can be installed in new *and* existing homes. As is often the case, it is much easier to incorporate a greywater planter into a new home — that is, a home that's under construction — than to retrofit an existing home. So, let's start with new construction.

Solar Design Tip

When designing a solar home that will house a built-in greywater planter, be sure not to install sloped glass on the south side of the home. Angled glass allows sunlight to enter year round but can lead to severe overheating in the summer. It's also difficult to cover with shades, so it can lose a great deal of heat in the winter. Moreover, sloped glass wall designs expose the glass to sunlight 12 months a year. The continual exposure to direct sunlight (heat) and then cooler nighttime air causes the seals between the double glass panes to expand and contract. Small cracks eventually form, allowing moisture in between the panes of glass. Moisture accumulating inside a double pane window looks ugly and can leave deposits on the inside surface glass when it evaporates. Moisture and mineral deposits also block sunlight and will necessitate costly replacement sometimes in only a few years.

Installing Greywater Planters in a New Home

If you're building a passive solar home, indoor greywater planters can be built along the south side of the home — for example, in the living room or in a hallway that runs along the south side of the home to bedrooms (Figure 8-2). Pipes can be run directly from sinks, showers, and washing machines allowing you to divert greywater directly to planters so there is no need to store greywater. (Remember, greywater gets pretty smelly, pretty fast, so it's best to use it immediately.)

Fig. 8-2: Greywater Planter Placement. *Greywater planters can be located indoors or in attached or detached greenhouses.*

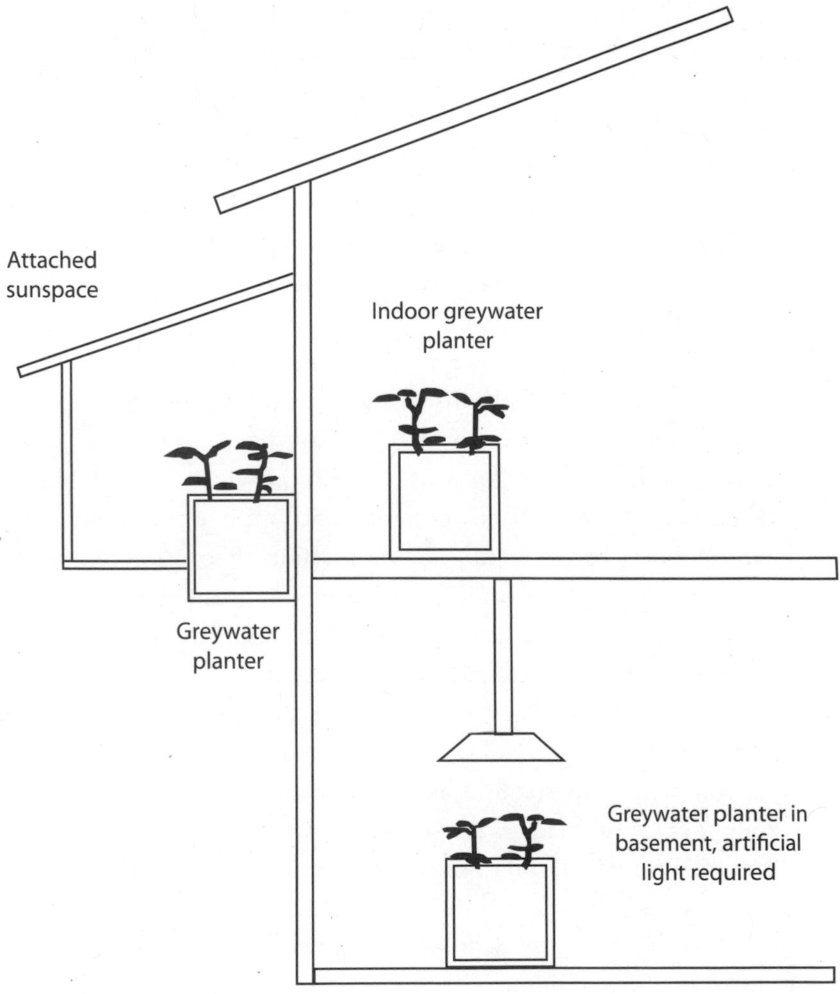

Designing a greywater system into a new home also makes it easier to ensure that greywater flows to planters by gravity. Avoiding pumps saves money, energy, and maintenance. Gravity flow greywater systems are the simplest and most environmentally sound.

If at all possible, avoid high-tech pump-driven greywater systems. High-tech systems are typically designed so that greywater empties into a central (temporary) storage tank, known as a surge tank (discussed in Chapter 6). The water is filtered and then distributed to planters with the aid of an electric pump.

While pump-driven systems will successfully capture water, they tend to defeat the central purpose of a greywater systems: they remove many of the nutrients we are trying to capture and put to good use.

My advice is to keep your system simple. Let water flow directly into planters by gravity and steer clear of storage tanks, pumps, and extensive systems of pipes.

Another option in new construction is to install greywater planters in a greenhouse built onto — or right next to — your new home. In the solar design world, we refer to greenhouses built onto homes as "attached sunspaces." Figure 8-2 shows this option.

If designed correctly, an attached sunspace (greenhouse) can provide additional heat for your home in the winter as well as food. Be careful, however. If not designed and built correctly, attached sunspaces can become a huge energy liability. Because greenhouses are merely glass boxes, they can lose tremendous amounts of heat in the winter. In the summer, they can gain lots of heat, causing your home to overheat. Before you design and build one, be sure to read the section on attached sunspaces in one of my most popular best-selling books, *The Solar House: Passive Heating and Cooling.* It provides sound advice on designing and building an energy-efficient attached sunspace.

As just noted, another option is to build a detached greenhouse near your new home. Although this structure probably won't help heat your home in the winter, if designed correctly, it could provide a significant amount of food for you and your family. Once again, design carefully. Avoid conventional greenhouses. They are an energy nightmare! I recommend energy-efficient Chinese greenhouses for year-round growing. To learn more, you might consider enrolling in one of my workshop on Chinese greenhouses at The Evergreen Institute or attending one of my off-grid aquaponics courses.

Installing Greywater Planters in an Existing Home

Installing an indoor greywater planter in an existing house can be extremely challenging. Placement of the planter is the biggest challenge you'll face. The easiest location in most homes is the basement because basements often provide easy access to drain pipes from sinks and showers. In the basement, you may also be able to isolate a greywater drain pipe from the blackwater pipe, then run the greywater directly into the greywater planter. (Details of this method are outlined in Chapter 6.) Unfortunately, a basement greywater planter will very likely require a lot of artificial lighting, making the system less environmentally sound.

Placing a greywater planter upstairs near a sunny window will very likely reduce but not eliminate the need for artificial lighting. This option, however, is not as easy as it sounds. The biggest challenge is finding free space near a window. The second challenge is routing greywater pipes from sinks and showers to a planter. Even if it is possible, you may need to hire a professional plumber to perform the work. He or she will most likely need to open up the walls to access drains from sinks and showers.

The best location for natural sunlight is near a window on the south side of your home, provided it receives plenty of winter sun. What you will soon discover, however, is that if your windows are protected by overhangs (eaves), very little sunlight enters through these windows during the summer. That's because properly sized eaves block the high-angled summer sun. This design feature helps shade windows during the hottest part of the year and, in so doing, reduces summertime solar gain. In the summer, most vegetables will have a devil of a time growing inside a home without supplemental lighting. Ironically, the optimal growing time for an indoor planter near a south-facing window is in the winter. That's when the low-angled winter sun penetrates windows. It warms your home and provides a great deal of visible light that supports luxuriant plant growth.

If you have a suitable location to install a greywater planter, you will need to either build one or buy one. Whatever you do, be sure it is watertight. Also be sure that you buy or build one with an automatic overflow — that is, pipes that prevent the system from filling up with greywater and overflowing, spilling greywater onto your floors. A drain pipe that runs from the planter to the blackwater line will suffice. Or, if you would like to capture the overflow, why not run it to an outdoor irrigation system like those discussed in Chapter 6?

If you lack space inside your home or don't have a good location, consider installing a greenhouse, either attached or detached, and install your greywater planters in it.

Now that you understand a little about your options, let's look at indoor greywater planters in more detail. This information will help you design a system to meet your needs.

Design and Construction of an Indoor Greywater Planter

I first became aware of indoor greywater planters in the 1980s when researching environmentally sustainable home building. It was then that I ran across the work of maverick designer and builder Michael Reynolds. His homes, called Earthships, (Figure 8-3) are solar-powered earth-sheltered structures whose walls are built out of hundreds of used automobile tires packed with dirt (subsoil). The interior walls are typically covered with a beautiful earthen plaster.

Designed for super-efficiency, off-grid solar living, Earthships capture rainwater from the roof. The water is stored in two large cisterns, then filtered before use. Rainwater supplies all household needs: cooking, bathing, and drinking.

Earthships are heated by the sun — using a technique known as passive solar. All the electricity is provided by small solar electric systems. Because these homes are so efficient, homeowners

Fig. 8-3: Earthship. *This amazing home provides many of its occupants' needs, including food produced in botanical cells—indoor greywater planters.*

require solar electric systems that are a fraction of the size of those required in conventional homes.

Earthships are also equipped with solar thermal systems that generate hot water for bathing, washing dishes, and so on. Sheltered by earth, these homes are also naturally cooled. They are built using locally available and abundant materials: dirt from the site and hundreds of used automobile tires.

The Earthship is one of the most — if not *the* most — sustainable and self-sufficient homes in the world, perhaps even in the Milky Way galaxy. Among the many features of these remarkably efficient solar-powered homes is a greywater system used to grow plants year round.

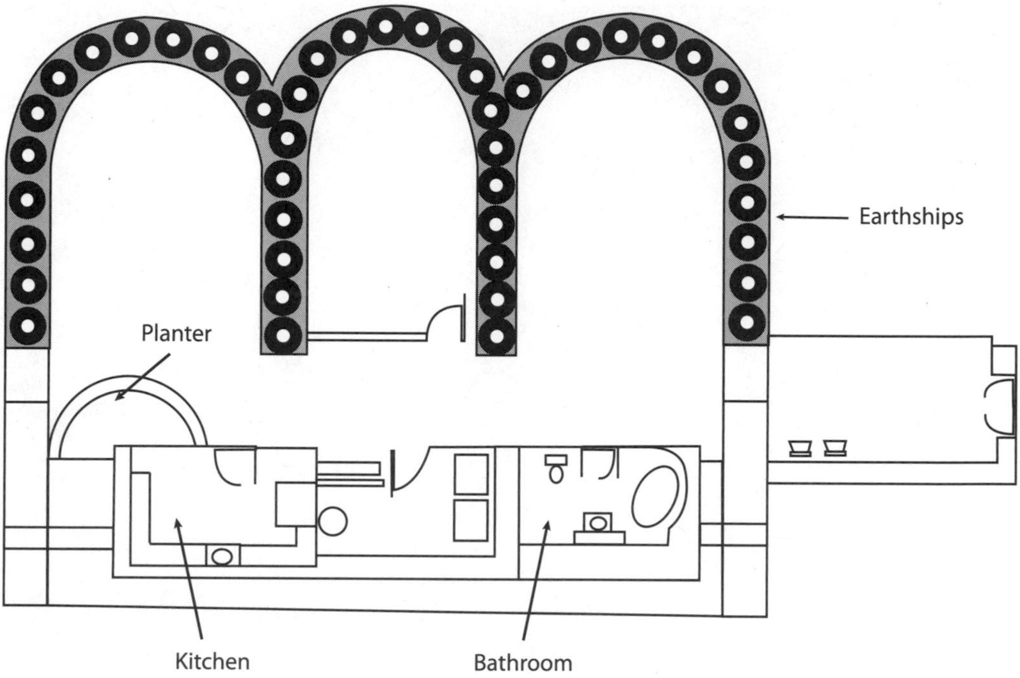

Fig. 8-4: Earthship Botanical Cells. *Greywater planters or botanical cells are designed similarly to subsurface-flow wetlands discussed in the previous chapter. They provide homeowners with an opportunity to grow food with "waste" water — specifically, greywater.*

As shown in Figure 8-4, greywater planters are strategically located so they receive greywater directly from sinks, showers, and tubs. In Earthships, this water comes from a rooftop rainwater catchment system. The water is typically piped from the roof to two storage tanks, called cisterns, one on the east and the other on the west side of the home. Rainwater flows from the cistern when needed through a filtration system, then to sinks, showers, tubs, and toilets. Greywater flows from sinks and the like by gravity into indoor greywater planters. It nourishes plants. In Earthships, the remaining filtered greywater is captured and used to flush toilets. (From toilets, it flows to a septic tank and then a lined outdoor planter.)

Figure 8-5 shows the anatomy of a greywater planter, referred to as a botanical cell by Reynolds and his colleagues and the hundreds of people who live in these earth-friendly homes. The walls of the botanical cells are built from poured concrete or cement blocks (mortared in place). Each planter is lined by a waterproof

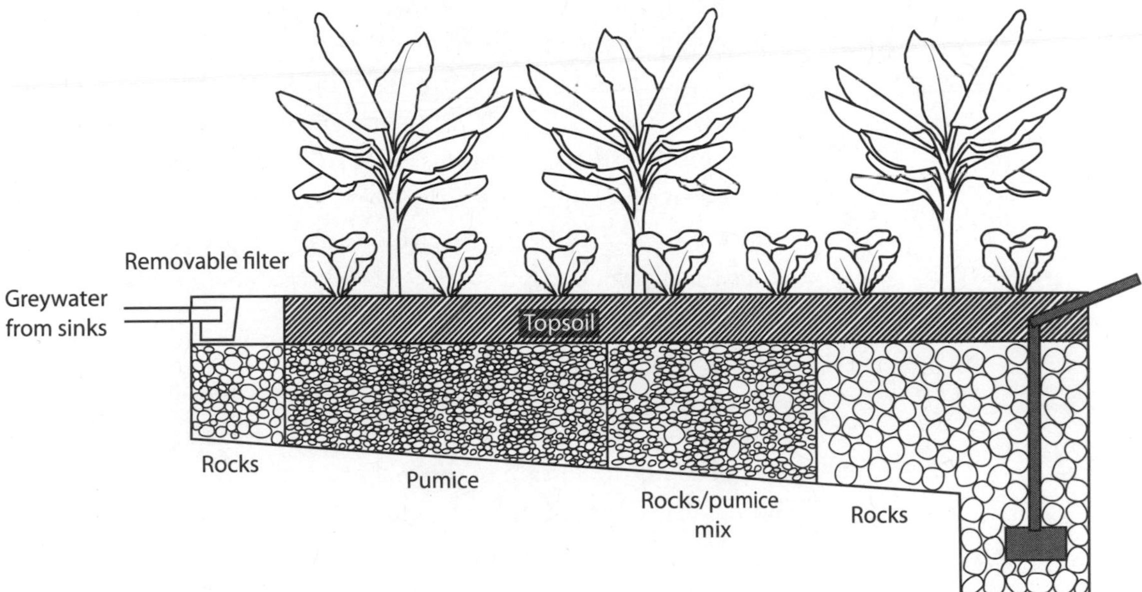

Fig. 8-5: Details of Indoor Botanical Cell. *Baffles separate the botanical cell into four compartments and increase the residence time of greywater in the system, helping to ensure complete biological cleansing.*

membrane. Pond liners made from EPDM or other types of rubber are commonly used. The planter is divided into four chambers by rubber partitions. Holes are cut in the partitions to allow water to flow from one compartment to the next. The bottom 12 to 18 inches (30 to 46 cm) of the planter is filled with lava rocks, which are common in New Mexico, or clean crushed rock like that used in submerged wetlands. Soil is placed over the surface, forming a layer 12 inches (30 cm) deep.

As illustrated in Figure 8-5, greywater from sinks, showers, and washing machines empties into the first chamber at the upper end of the planter. It first flows into and through a removable filter that snags hair, lint, grease, oils, and other large particles. (This prevents them from clogging up the rock filtration bed. The grease filter must be emptied and cleaned from time to time. The frequency of cleaning depends on the quantity and quality of one's greywater.)

Greywater then flows into the rock bed of the first chamber. It proceeds along the length of the first chamber by gravity, flowing through the rocks beneath the soil, then into the second chamber through an opening in the partition between the two. Holes are cut on opposite ends of adjacent partitions. This forces water entering one end of a chamber to flow through the entire rock bed of that chamber before draining into the next one.

In this design, the partitions act as baffles. They slow the flow of greywater through planter. This increases residence time of greywater in each chamber and, as you read in the previous chapter on constructed wetlands, it therefore enhances biological decomposition of organic materials.

Microbial decomposition of dissolved and suspended organic molecules begins as soon as greywater enters the system. As greywater flows from one chamber to the next, bacteria and other microbes on the surface of the rocks and roots of plants that grow down into the rock bed, break down organic matter. The water becomes progressively cleaner as it progresses through the system. The roots of plants in the rock bed absorb water and nutrients.

Greywater planters can be used to grow vegetables such as lettuce, tomatoes, Swiss chard, New Zealand spinach, and others. You can even grow bananas, avocados, and dwarf fruit trees in Earthship-inspired botanical cells.

As noted earlier, not all of the water flowing through the system will be used by the plants, especially if the planter services a high-volume source of greywater or the planter is undersized for its source. To accommodate the excess, the purified greywater, the fourth chamber in the botanical cell contains a small depression or well. A submersible electric pump placed in the bottom of the well pumps water out. It is controlled by an automatic float switch.

In Earthships, purified greywater is then pumped out to a series of filters that remove any remaining "impurities." Although the filtered water is not drinkable at this stage, it is clean enough for other uses — notably, flushing toilets. The water could also be used to water outdoor plants or plants growing in a greenhouse. If you are interested in installing a water purification system like this, don't sweat. You can order a pre-plumbed pump and filter kit — known as a water organizing module — from the folks at Earthship Biotecture.

Purified greywater that is used to flush toilets becomes blackwater. Not to waste this water or the nutrients it contains, Reynolds' Earthships pipe the blackwater to a solar-heated septic tank. Located on the south side of the home, the septic tank precipitates out solids — just like a conventional septic tank. The liquid draining from the tank then flows to an outdoor planter — or several planters — known as outdoor botanical cells.

Outdoor botanical cells are lined planters just like indoor planters with a rock base for filtration and soil top for rooting plants. Nutrients and water from the solar septic tank feed vegetables and flowers.

Over the years, I have visited numerous Earthships. One I visited in the mountains of Colorado (near Telluride) had the first (I think) outdoor botanical cell. Interestingly, the effluent from the botanical cell was much cleaner than the treated effluent from the town's more technologically sophisticated and costly sewage treatment plant. Engineers from the sewage treatment facility were as amazed as I was and, unlike me, a bit skeptical. (As a side note, to satisfy skeptical code officials, the septic tank may need to be connected to a conventional leach field as a backup.)

Botanical cells work, and they work well. In fact, the Earthship greywater system has been tweaked by Reynolds and his colleagues for several decades. And, what is more, they create a lovely, productive garden inside your home — or in a greenhouse. For more details, be sure to read *Greywater: Containment, Treatment, and Distribution Systems.* It's part of the Earthship Chronicles series published by Reynolds and his colleagues. Or schedule a visit to Earthship Biotecture headquarters in Taos, New Mexico. It's well worth the trip!

Combined with rainwater catchment systems, greywater planters allow environmentally conscious homeowners to make the most out of every drop of water they use. They help us minimize our impact on planet Earth, which, lest we forget, is the only habitable piece of real estate that's been found in the Universe — so far, at least. Far from being a detriment to the future of life on Earth, you can become a positive force for good by installing a greywater planter.

Far from being a detriment to the future of life on Earth, you can become a positive force for good by installing a greywater planter.

Indoor Greywater Planters in Existing Homes

If you're thinking about installing a greywater planter in an existing home and you have the space to accommodate an indoor planter, you'll find that a botanical cell adds beauty to your home — and can produce a significant amount of food — if it receives plenty of light, preferably natural sunlight.

Figure 8-6 shows the components of a greywater planter for an existing home. As you can see, it's like the botanical cells of Earthships, but differs in several key aspects.

As shown in Figure 8-6, in these systems, greywater is first filtered to remove large particles and fibers that could clog up the works. The filtered water is then piped to a soil box planter. The water then flows through infiltration pipes — porous pipes in the soil that allow the greywater to seep into the soil without clogging the openings in the pipe.

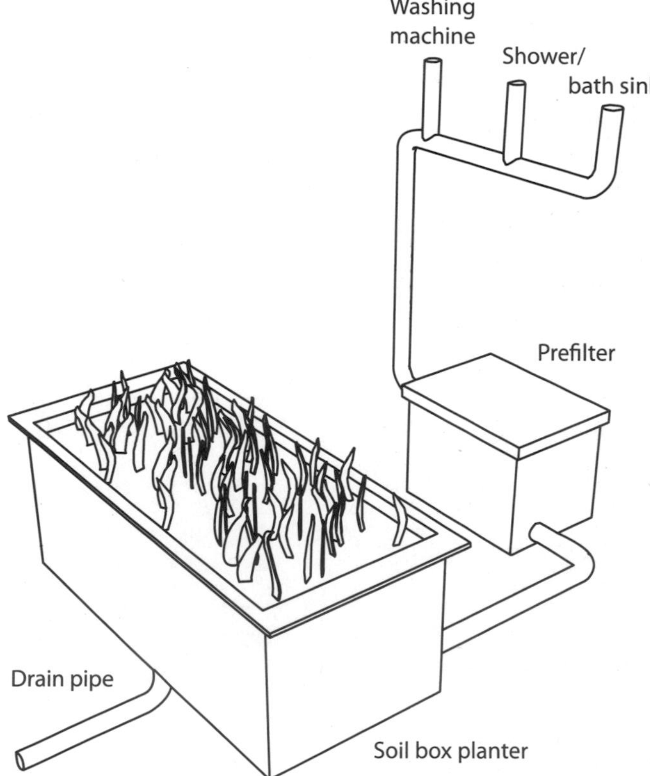

Fig. 8-6: Indoor Greywater Planters. *This system consists of a planter and a prefilter. The system is like a botanical cell but pre-treats greywater, thus reducing the system's overall utility.*

Buying an indoor greywater filter may be challenging. I'm only aware of two commercially available greywater filters, although there may be more systems out there — especially in countries like Australia where greywater recycling is popular. The two systems I'm aware of are Aqua2User and Envirolet's Greywater Biolan 125 and 70. Both are designed for filtering greywater that's used outdoors for irrigation. Both come equipped with a series of filters that screen out gunk and particulates. Bacteria and other microorganisms growing on the filters consume the organic matter.

I have no experience with either of these system, so can't vouch for their effectiveness. However, I suspect that they could be used to supply filtered water to an indoor planter.

You may want to build your own filtration system. Although I haven't built an indoor greywater filter, I have built and tested several outdoor pond filters that remove algae and duck poo from several small artificial ponds. (You can see one online at YouTube, "Biological Pond Filter at Evergreen Institute.")

My pond filters have proved to be extremely successful for 200-gallon ponds. A photo of one filtration system that I installed in 2014 is shown in Figure 8-7. In this system, nutrient- and algae-rich water is pumped out of the pool into a 55-gallon (about 200 liter) plastic drum. Water flows through a dozen scrub pads held in a metal mesh minnow "bucket." The water flows from here through my grow beds.

Fig. 8-7: Pool Filter. *This filter in this plastic barrel contains scrub pads that filter out huge amounts of algae and duck waste very rapidly. Bacteria and other microorganisms break down the organic matter. The filtered water full of nutrients then flows into four plastic totes, two of which are used to grow algae and duckweed, and two of which are used to grow vegetables.*

Although I cannot vouch for its effectiveness as an indoor greywater system prefilter, my duck pool filter does a great job cleaning up some rather yucky water rich with algae and duck poo. If it cleans the water in their 200-gallon pond — which gets pretty dirty, pretty fast — it should do a good job prefiltering household greywater that can then be pumped into your indoor greywater planter.

Figure 8-8 shows a design for another indoor greywater planter based on another outdoor biological pond filter of mine. As illustrated, in this system greywater first flows into gravel filter/growing bed — a plastic tote filled with pea gravel. Effluent then flows into a soil box/growing bed. Be sure the prefilter and soil box are 100% waterproof containers.

As shown in Figure 8-8, the bottom of the soil box is filled with crushed rock or pea gravel, as in the Earthship botanical cell. Coarse sand is layered on top of the rock. It is separated from the rock layer by a window screen or geotextile fabric. Finer sand is layered on top of that. The top 2 feet consists of an organic-rich topsoil — humus is ideal! Be sure not to use soil containing a lot of clay, as clay particles can become dislodged from the soil and "trickle" down into the underlying sand and rock.

As illustrated in Figure 8-6, a drain pipe in the bottom of the box allows excess water to be removed. That's important so the

Fig. 8-8: Proposed Design for Indoor Greywater Filter. *A tub filled with gravel filters out organic and inorganic wastes that nourish plants (not shown here). Partially filtered water then flows into a greywater planter to complete purification and grow additional vegetables.*

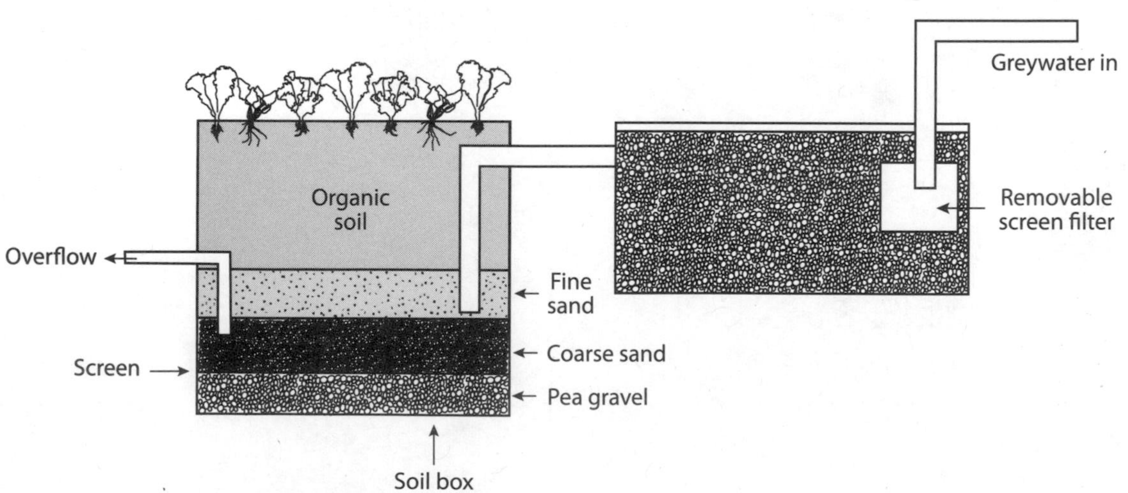

water doesn't saturate the soil and kill your plants. As all good gardeners know, when the soil becomes saturated with water, tiny air spaces between the soil particles fill up, roots will suffocate from lack of oxygen, and plants will die.

When depositing greywater directly into soil planters, be sure to install pressure infiltration pipes, shown in Figure 8-9. As illustrated, pressure infiltration pipes consists of a perforated pipe with holes drilled along its length. A sleeve is affixed to the upper half of the pipe. The pipes are buried in a 2-to-4-inch (5 to 10 cm) layer of wood chips or mulch layered on top of the top soil. Water flowing through the perforated pipe forces the outer sleeve to open, allowing water to stream out. When water flow stops, the sleeve settles back onto the perforated pipe, blocking the holes and preventing dirt, worms, and roots from entering and clogging the works.

Water draining from the soil box flows into another, smaller tank that contains a submersible pump. It pumps the water outdoors to trees, shrubs, or other plants. This pump is controlled by a float switch to prevent overflow.

For best results and the least amount of maintenance, it's best to use greywater from bathroom sinks, showers, and tubs, not kitchen sinks. That's because kitchen sinks can introduce a lot of organic matter — especially if you use a garbage disposal — that will require frequent cleaning of filters.

Greywater planters can be installed indoors and outdoors. The biggest challenge in an outdoor greywater planter is preventing

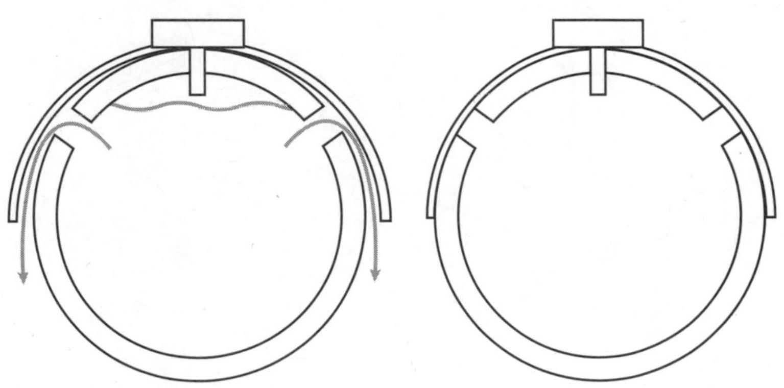

Fig. 8-9: Infiltration Pipe. *This pipe feeds water into a layer of mulch on the surface of a soil-based greywater planter. Water pressure forces water out of pipe (on left).*

heavy rains from flooding the planter. Be sure to install a well and a submersible pump on all outdoor planters to prevent this problem. Also, be sure to pump the overflow to a location where it will immediately percolate into the ground — for instance, a large infiltration bed, such as those discussed in Chapter 7 and shown in Figure 7-8.

Watson-Wick Filters

Another option for treating greywater outdoors — and capturing its nutrients — is the Watson-Wick filter (Figure 8-10). I learned about this invention in the late 1990s, and helped install one at the Lama Foundation in northern New Mexico.

The invention of New Mexico's Tom Watson, this system consists of a bed of rock, such as pumice (lava rock) or clean crushed rock (like granite, not limestone). The rock is placed in an excavated depression much like a submerged wetland. The sides of the pit are bermed with dirt excavated from the site.

As shown in Figure 8-10, greywater enters through a pipe and flows into a cavity formed by a device called an infiltrator. This large, plastic C-shaped infiltrator creates a chamber where greywater enters. (You can purchase an infiltrator online.) Greywater

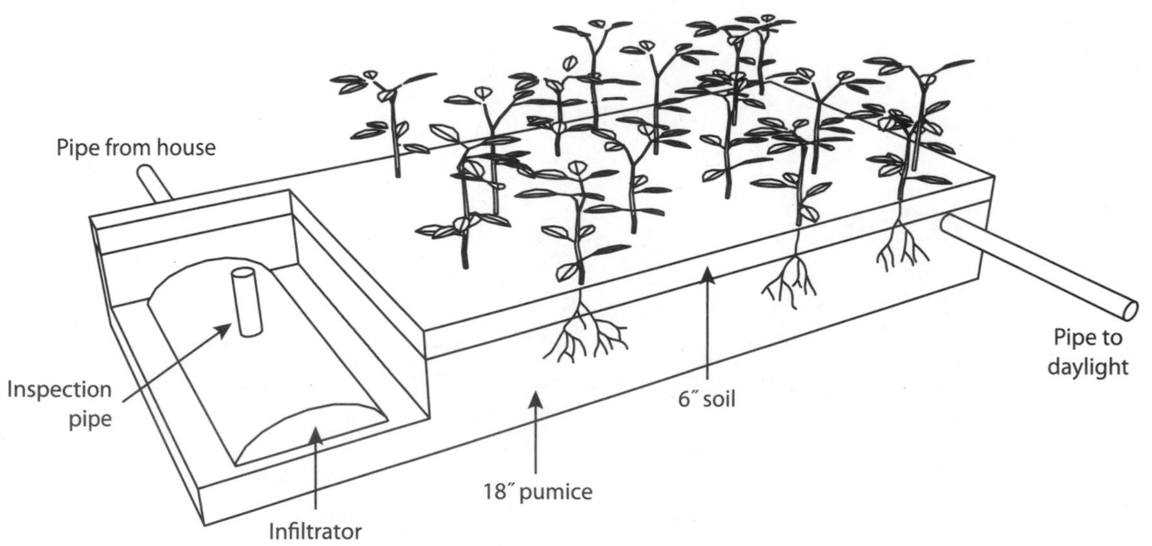

Fig. 8-10: Watson-Wick Filter. *This system, which resembles many other biological treatment centers, can be used to "treat" greywater and blackwater.*

flows by gravity along the length of the infiltrator. Microbes begin digesting the organic matter. Greywater also seeps out laterally through the adjoining rock bed. Microbes on the surface of the rocks and on the surface of roots "process" the organic wastes, liberating nutrients. Over the top of the rock bed is a layer of topsoil. It supports vegetables, flowers, berry bushes, and fruit trees. Their roots draw nutrients from the soil and the greywater flowing through the underlying rock bed. Purified water, if any, can be piped out of the bed to irrigate additional plants.

Hundreds of Watson-Wick filters have been installed in New Mexico. Like submerged wetlands, these simple biological filters can operate year round — if sized correctly. I'd strongly suggest you get more details online before building one and contact individuals who have installed them to learn do's and don'ts.

Conclusion

Greywater planters are small, contained subsurface-flow wetlands. They are yet another option for harvesting water and nutrients in the vast quantities of greywater we produce in our homes and places of work. Indoor planters designed into a house from the start add beauty and charm and, if done correctly, a year-round supply of vegetables, flowers, and possibly even some types of fruit such as lemons, limes, oranges, or bananas. Pay close attention to lighting. If artificial lighting is required, consider installing a solar electric system to provide the electricity needed to power the lights.

Afterword: Achieving Total Self-sufficiency

If you are like most of my readers, you are somewhere on the path to a self-sufficient lifestyle. While reasons for this pursuit vary widely — from fear that our economic system will collapse to global catastrophe caused by climate change — the steps to self-reliance are pretty much the same. This book has outlined one of the key steps on the path: capturing the valuable nutrients in urine, humanure, and greywater, and then returning these nutrients to soil to grow food and perhaps even food for poultry or livestock that, in turn, provide food and other valuable resources for us.

I've spent a lifetime studying, experimenting with, and writing about self-sufficiency and sustainability. In this afterword, I wanted to share some of my ideas on the topic with you. In this section, I will outline key measures you, your family, and friends can and must take to create a self-sufficient lifestyle. To make it easier, I've broken the task down into monthly goals. If you follow my plan, you could be well on your way to total self-reliance in a year and a half. Some of you may have already completed some of the tasks, so the time it will take to reach your final goal will be shorter. For those who are just starting out, there's no reason you can't pursue several ideas at once, accelerating the process by many months.

I'd recommend that all readers implement some of the goals concurrently. More specifically, I'd recommend starting with the goals set out in months 11–13 which have to do with health, fitness, and financial security. Read all these ideas now and start implementing them as you develop energy, food, and water self-sufficiency.

Month One: Energy Self-sufficiency — Seal Up the Leaks, Now!

We need energy to live — to enjoy our lives, make a living, and accomplish our goals. Self-sufficiency and sustainable living can be achieved in part by a drastic reduction in energy use.

Sadly, Americans, Canadians, and many others in the more developed nations waste most of the energy they consume. The silver lining behind that cloud is that our waste represents a huge untapped resource.

To reach your goals of self-sufficiency, one thing you and your loved ones can do is to find ways to dramatically reduce energy waste — to become more efficient. Fortunately, there are many simple and cost-effective measures we can each take to trim the fat from our energy demand.

Over the years, I've found that most people think they are frugal. They brag of $100 to $200 per month utility bills. An energy-efficient household can function on a fraction of that energy. A self-sufficient household must.

The most important measure you can take to reduce your waste is one of the least expensive and easiest: sealing up the many energy-robbing leaks in the building envelope of your home. Get on the phone now and call an energy auditor. Ask him or her to come to your home to perform an energy audit, including a blower door test.

Energy audits will help you pinpoint the many ways you are currently wasting precious and costly energy. The blower door will reveal how leaky your home is — that is, how much cold air leaks in during the winter and how much hot air leaks in during the summer. The blower door test allows you to identify the leaks so they can be sealed.

Once the audit has been performed, seal up all the leaks that are discovered. Caulk and weather stripping will be your major allies in this effort. They're cheap and easy to apply. You can do the work yourself or hire a professional to do it. Be sure to communicate your desire for total energy self-sufficiency to your auditor and energy retrofitter so he or she knows you are very serious. Let them know that you'd like to cut energy demand by 75%!

Sealing the leaks in the building envelope will dramatically reduce your heating and cooling costs as well as your dependence on energy companies and local utilities. It is one of *the* least expensive and most cost effective measures you can take to create a more energy-self-sufficient and sustainable lifestyle. I won't detail all the places you'll find leaks or ways to seal them. You can learn more about this in one of the many books I've written on energy like *The Solar House, Power from the Wind,* or *Green Home Improvement.* The latter book is especially helpful. It contains dozens of home energy-efficiency projects you can complete to trim the fat.

Month Two: Insulate the Hell Out of Your Home

The next step on your path to energy-self-sufficiency is to beef up the insulation in your attic and walls and under crawlspaces and around foundations. Even if you have already added insulation, you'll find that adding more will often help. Most insulation retrofits simply bring buildings up to code, but most building codes are woefully insufficient when it comes to insulation. Today, smart home energy experts are calling for much higher levels of insulation than code officials do. (More on this shortly.)

Insulation can begin as soon as you've sealed up the leaks in your home, so it can be achieved in the first month, if you would like. Whatever you do, don't skip step one. That is, do not, under any circumstances, insulate your home until you have sealed up all the leaks.

Air sealing must precede insulation because air leaking through walls carries moisture with it. Moisture is deposited in the insulation in wall cavities and ceilings, and even a tiny bit of moisture in most forms of insulation will reduce the insulating value (R-value) of a wall or ceiling by half. Wet insulation in a home is like wearing a wet shirt on a cold day. It doesn't hold heat in. In fact, moisture actually helps heat escape.

So, once your house is sealed tightly, insulate the hell out of it! How much insulation do you need?

To make your home as energy efficient as possible and to dramatically reduce your energy demand, I recommend insulating ceilings to R-50 to R-80 in all climate zones — hot and cold — and insulating walls to R-30 to R-40. If your home is built over a crawl space or rooms are built over an unheated garage, insulation under the floor should be at least R-25.

Combined with air sealing, high levels of insulation could easily slash heating and cooling costs by 30% to 50%. It all depends on how leaky and how poorly insulated your home was at the outset. If your home is extremely leaky and very poorly insulated, you might even see a decrease in heating and cooling costs by 50% to 75%. Don't let insulators talk you into lower amounts of insulation. Your goal is total self-sufficiency, and you want to offset any settling of blown-in insulation (like cellulose) and the accumulation of moisture — both of which will lower the R-value.

To fully insulate your home, you will also need to install insulated shades over windows — and use them. Another option is to place rigid foam insulation panels in your windows at night on cold winter nights to hold the heat in. Rigid foam inserts can be covered with fabric so they look nice. In my experience, foam insulation panels in windows can raise room temperature dramatically in small rooms — about 10° to 15°F — and thus dramatically increase comfort.

Month Three: Install Energy-efficient Lighting, Appliances, and Electronics

Once you've sealed your home, "supersized" its insulation, and insulated windows, the next step toward energy self-sufficiency is to trim plug loads. *Plug loads* refers to devices like toasters, coffee pots, and TVs that draw power from a plug inserted in an electrical outlet. Reducing plug loads can be achieved by eliminating unnecessary appliances or replacing them with manually operated models. Replace the electric knife with a serrated knife. Get rid of the electric toothbrush. Dump that electric mixer and use a manually operated egg-beater — or simply use the latter most of the time.

The same recommendation applies to your family's clothes dryer. Either replace it with a solar clothes dryer — aka a clothesline — or keep the dryer, but use it sparingly.

Consider retiring your automatic dishwasher, too. If you are careful, you can wash dishes by hand and use less water than the most energy- and water-efficient dishwashers. I've washed dishes forever using two tubs — one for hot soapy water, the other for clean rinse water. Use the rinse water to irrigate plants inside and outside your home.

While you're trimming your energy demand, get rid of extra freezers and refrigerators — especially old fridges and freezers in your basement or garage to hold extra food and drinks. Older model freezers and refrigerators are energy hogs that are robbing you blind. They often use three to four times more energy than their modern counterparts.

While you're at it, replace energy-inefficient incandescent lights with the more energy-efficient compact fluorescents or LEDs. LEDs are now more efficient than CFLs and last a lot longer — although they cost 10 to 20 times more. I'd suggest installing the longer-lasting bulbs wherever you can. When in a self-sufficiency mode, durability and efficiency rule!

Another measure to consider is installing a programmable thermostat. If set properly, it can easily reduce energy demand by 10%. Programmable thermostats are vital to those who are gone during much of the day.

Month Four: Energy Self-sufficiency — Renewable Energy to Heat Your Home

Now that you've trimmed your energy demand by sealing leaks in the building envelope of your home, upgrading insulation, and trimming plug loads, you should be well along the path to complete energy self-sufficiency. If you have been diligent, you could easily have reduced your household energy demand by 75%! In the process, you've lowered your cost of living and the number of hours you need to work each day to get by. And, lest we forget, you've also made your home a lot more comfortable — warmer in the winter and cooler in the summer.

Now it's time to take the final step toward total energy self-sufficiency: generating the energy you need. This is a huge challenge that could take several months to complete it. If you are eager to achieve energy self-sufficiency, however, you could hire someone to do this work and have it completed within a month. Let's begin with ways to heat your home.

There are many ways to heat a home using renewable energy. One option is wood. If you live in an area with ample wood supply, consider installing one or two efficient, clean-burning woodstoves. Woodstoves can be supplied with wood from forests or from wood you collect locally — for example, from tree trimmers. Tree trimmers often generate an abundance of waste wood that they truck to the landfill. Used wood pallets can be another valuable source of firewood. Many companies receive shipments on wood pallets. They often stockpile them, then discard them. Look around. When you find a supply, ask if you can have them. If you don't own a truck, consider buying a small trailer for your car to transport them to your home.

Your neighbors may also be a valuable source of wood. Many trim trees and dump the "wastes." They also discard leftover wood — like 2 x 4s after home improvement projects. Ask around and you'll very likely find several people in your neighborhood who will donate their "waste" wood.

Another option for suburban dwellers is to create a small woodlot in your backyard. Plant fast-growing trees like poplars that can be harvested, cut into cordwood, and stockpiled in a dry location for hard times. If you live in the country, you may already have access to a small woodlot, or you can plant one yourself in fields around your home. If you have insulated and sealed your home, chances are you won't need a huge supply of wood to stay warm in the winter.

Wood is a renewable energy resource but, to be honest, it is the dirtiest of all renewable energy options. A cleaner and more sustainable way to provide wintertime heat is by incorporating passive solar heat in your home.

In a passive solar home, sunlight from the low-angled winter sun streams through south-facing windows, heating the interior of a home. If your home has numerous south-facing windows, you've probably already noticed the natural wintertime heating. You may be able to install a few more south-facing windows in rooms that require heat.

Another option to heat your home passively is to install a solar greenhouse on the south side of your home, also known as an attached sunspace. If properly designed, a solar greenhouse can provide a substantial amount of solar heat in the winter, thus aiding your efforts to take care of your own energy needs, and it can be used to grow food with greywater. I briefly discussed attached sunspaces in Chapter 8.

Passive solar heating requires a great deal of knowledge, so learn everything you can before you start. As many readers may know, I started my solar energy career in passive design and have consulted on dozens of homes throughout North America. I've also written several articles on incorporating passive solar into existing homes in various magazines like *Home Power.* My most thorough treatment, however, is in my book, *The Solar House.* You can also learn more about this topic in *Solar Home Heating Basics.* This book describes other solar options as well, such as solar thermal and solar hot air systems. Finally, it's worth noting that I teach numerous workshops on passive solar heating at The Evergreen Institute and through Mother Earth News Fairs held in various parts of the country.

Yet another option to solar heat your home is a technology called solar hot air. It requires solar collectors mounted on the south side of your home — either on a rack next to your home or on south-facing walls. Indoor air circulates through the collectors when the sun's shining. It is heated by the sun, then circulated through the adjoining room (or rooms) they supply.

Month Five: Energy Self-sufficiency — Solar Hot Water System

Now that you've taken steps to make your home super-efficient and heat your home with renewables, you'll need to create a way to heat water for cooking, washing, and bathing. If you've installed a woodstove, it can be used to cook and to boil water for tea.

You may also want to consider installing a solar hot water system. Also known as solar thermal systems, these systems last forever — if installed right — and can provide the bulk of your hot water in warmer, sunnier regions of the world. In the southern United States and in tropical countries, for instance, it's possible to produce nearly all of your hot water using these systems. In cooler, less sunny climates like the Midwest, it's possible to meet to 70% to 80% of your hot water needs.

In recent years, some companies have begun installing hybrid water heaters powered by solar electricity. Hybrid water heaters contains an electric heating element and a device known as heat pump, more precisely an air-source heat pump.

Air-source heat pumps are ingenious devices that extract heat from the air in a home and use it to heat water inside the tank. They're extremely efficient. The backup heating element provides the rest.

Powering a hybrid water heater with a small solar electric system is often less expensive than a conventional solar hot water system for domestic hot water. It all depends on installation costs. If an installer is charging $10,000 to install a solar hot water system, a hybrid water heater powered by a few solar electric modules will be much more economical. Consider this option very carefully.

Also keep in mind that more frugal use of hot water will also help meet your demands for hot water. Be sure to install super water- and energy-efficient showerheads and flow restrictors on faucets. Also insulate all the hot water pipes in your home. Doing so will help you meet your needs for hot water more economically. If your existing water heater is not insulated, wrap it with a water heater jacket or blanket.

You may also want to consider installing an on-demand water heater. They can cut your energy consumption for hot water by 10% to 35%, depending on the volume of use.

Month Six: Natural, Self-sufficient Cooling

In many regions of the world, cooling is vital to comfort and survival. In fact, cooling can be a more significant challenge than heating homes in hot climates. To achieve self-sufficiency, then, you will need to find low-energy ways to efficiently cool your home. Fortunately, the air-sealing and insulation projects you completed to make your home warmer in the winter are also vital to keeping it cool in the summer, as are energy-efficient lighting and energy-efficient appliances. Less waste heat from these and other electronic devices — including lightbulbs — will help keep your home cooler through the dog days of summer.

You can achieve additional — and rather dramatic — reductions in energy demand to cool your home by painting your home a lighter color. A fresh coat of white paint reflects more sunlight in the summer and will naturally keep you and your family much cooler. A lighter color roof will help, as well. If you need a new roof, consider installing a lighter color metal roof. A layer of insulation beneath the metal roofing will also help keep your home much cooler in the summer.

Planting a few trees around your home to provide summer shade will pay off handsomely as well. Be sure you don't shade the south side of your home and reduce passive solar gain or shade any solar electric modules or solar hot water collectors.

Further measures that promote low-energy use and self-reliance include installation and use of ceiling fans, whole house fans, and solar-powered attic fans. Combined with air sealing, insulation, use of window shades, and energy-efficient appliances, these measures may nearly eliminate the need for air conditioning, making your home a model of energy self-sufficiency.

I discuss numerous passive cooling measures in my book, *The Solar House and Green Home Inprovement.* I also teach frequent workshops on the subject through my educational center and seminars at the Mother Earth News Fairs.

Month Seven: Energy Self-sufficiency — Solar Electric System

To achieve total energy self-sufficiency, you will also need to generate electricity. Fortunately, if you have taken all of the previous steps, you won't need much.

Solar electric systems are ideal for those whose roofs or yards are unshaded throughout the year. If properly sized and properly installed, a solar electric system can meet all the electrical energy needs of an energy-efficient household. If you live in a windy rural area, such as western Kansas or eastern Wyoming, consider installing a wind turbine.

Most solar electric and wind energy systems installed nowadays are grid-tied. They produce electricity to supply homes and businesses. When producing a surplus, however, the excess is backfed onto the electrical grid. Customers' meters run backward. The utility credits each customer for the electricity he or she feeds into the utility network. They don't pay you for it directly, but let you have it back for free within an allotted time.

Some companies reconcile surpluses monthly. In other words, if your system generates a surplus one day, you can have that electricity back free of charge during that billing period. Other states have adopted annual net metering, which means customers can carry their surplus for up to one year. If a customer needs electricity at night or during cloudy periods, he or she can have it back, free of charge any time during the year — even ten months later!

Solar electric systems can provide 100% of a family's electrical needs. There is one problem, however, that you should be aware of. That is, if the electrical grid "goes down," grid-tied solar electric systems shut down. Even if it's a bright and sunny day, virtually all grid-tied solar electric systems shut down if the utility experiences a blackout — for example, if lightning has wiped out a transformer.

Automatic shutdown in case of power outage may seem ludicrous, but it serves a good reason: it protects line workers from being electrocuted by current being backfed onto an otherwise de-energized system while they are carrying out their repairs.

For true self-sufficiency, then, you might consider going off-grid. That is, you might consider cutting your ties to the electrical grid. To do so, you'll need to install a battery-based off-grid solar electric system — also known as a stand-alone system. While this option may seem appealing, consider some of the realities of off-grid systems. First, batteries are expensive. Secondly, batteries require careful monitoring and periodic maintenance (filling them with distilled water). For optimum performance, batteries must be housed in a special vented room that remains between 50° and

80°F (10° to 27°C) year round. Batteries will need to be replaced every 7 to 15 years, depending on how well you treat them.

Most important of all, the best solar batteries on the market hold very little electricity. If you are planning on going entirely off grid, you'll need to have trimmed your electrical demand by close to 90%. Forget about having an electric stove, electric water heater, and a conventional hot tub — unless you want to install a massive solar electric system and battery bank, both of which can be extremely costly. To learn more about battery-based solar systems, you may want to take a look at my book, *Power from the Sun* or sign up for some of our solar electric workshops.

Don't give up on grid-tied solar yet, however. One manufacturer has come up with an innovative inverter design that allows a homeowner to power some loads during power outages. The company is SMA. Their newest inverter can be wired to a dedicated electrical outlet that remains energized after the grid goes down. This outlet can be used to power refrigerators and freezers and a few other critical loads, but only when the sun is shining. When the sun goes down, you are out of luck. Check out SMA's Sunny Boy TL-US series inverters.

Month Eight: Water Self-sufficiency

Energy is so important that it can easily take seven months to a year to implement the measures needed to reduce one's dependence on outside energy and create a renewable energy supply that's all yours. Now that you've made great strides toward energy efficiency, let's turn to another key requirement for survival and self-sufficiency: water — water to drink, cook, wash, bathe with, grow vegetables, and raise poultry and lvestock.

If you have a well and have installed an off-grid solar electric system, you should be fine. It's still not a bad idea to develop other sources of water to reduce well pump run time and save energy.

If you depend on water from the city or town you live in, however, you may want to devise an alternative plan to meet your needs for water. The easiest way to supply fresh water is to install a rooftop rainwater catchment system. Water collected from garage, shed, and house roofs can be used for watering gardens that supply your food. As discussed in Chapter 6, greywater from washing machines, sinks, and showers can provide additional water for growing food.

Rain catchment systems are relatively easy to install. If your home is equipped with gutters and downspouts, install a system of underground pipes into which the downspouts drain. Run this pipe to a cistern, preferably an underground storage tank. (That way it won't freeze in the winter and won't be overrun with algae.)

If your home is not equipped with gutters and downspouts, get them installed now. Then add underground pipe and a cistern.

The size of a cistern can vary, depending on rainfall amounts and water use. You'll very likely need to install a 5,000- to 10,000-gallon tank to meet your needs. Be sure to use water efficiently to reduce the size of the tank you will need.

Install a small submersible AC or DC pump in the cistern to draw water out of the tank and deliver it to the house or to your garden or poultry and livestock, if any. We run some of our rainwater directly from various roofs into ponds for our ducks and stock tanks for our cattle.

If you intend to use rainwater for showering, washing dishes, and cooking, consider installing a *roof washer.* This is a device that prevents the initial slug of roof water from entering the cistern. This water may contain dirt, leaves, and bird droppings and should be kept out of the system. Once the roof is washed, however, the remaining rainfall is automatically delivered to the cistern.

Rainwater can be used unfiltered for watering fruit trees and vegetables. Rainwater can also be purified and used to flush toilets. It can even be purified enough for drinking and cooking.

To purify rainwater, first run it through a filter that removes dirt. For drinking water, next run rainwater through a more advanced filter system that removes microorganisms and pollutants. Carbon filters can remove potential toxicants present in rainfall.

For the cleanest water, consider replacing your asphalt shingle roof with a metal roof. Metal roofs last a lifetime and produce the cleanest water — far better than water from a shingled roof.

Month Nine: Food Self-sufficiency

Now that you've ensured a steady supply of energy and water, you need to turn your attention to yet another essential requirement: food. You need to grow most or all of your food.

My advice to urban and suburban dwellers is "to stop mowing and start growing." Take a shovel to your sod and gradually turn that beautiful green, but otherwise fairly useless, lawn of yours into edible landscape.

To prevent neighbors from complaining, begin slowly. First, plant some fruit trees. (Remember that most fruit trees need to be planted in groups of two or more for pollination.) Then convert a small section of the lawn to flowers and herbs sprinkled with lettuce, beets, Swiss chard, and carrots. Your nosy neighbors will never know what you're up to!

If you have a sunny, spacious backyard, plant a large garden and lots of fruit trees. Then learn how to freeze, can, and store (in root cellars) the products of your garden and mini orchard for use in the winter months. Consider teaming up with friends and neighbors to share surpluses. As any seasoned gardener will tell you, it is not unusual to produce more than you can consume, freeze, or can. One year, for example your summer squash will grow like weeds, but your tomatoes will mysteriously wilt. Meanwhile, a fellow gardener enjoys her best tomato crop ever, but her squash succumbed to squash bugs. Set up a network to trade surpluses so you can stockpile a diverse food supply. Can and freeze together to share the burden and build networks of interdependency, which, ironically, are vital for self-sufficiency.

Month Ten: Grow All Winter Long

Freezing, canning, and storing products of your garden can help you feed yourself and your family all year on a summer's bounty. It may provide all the fruits and vegetables you need.

If you want to eat fresh lettuce or spinach in the winter, however, consider installing a greenhouse. Because traditional greenhouses are notoriously energy inefficient, consider installing a four-season greenhouse, that is, a hoop house with smaller hoop "houses" perched over raised beds.

This system, if designed and managed properly, will allow you to continue to grow cold-footed plants such as spinach, lettuce, bok choi, and others throughout the coldest of winters. Check out Eliot Coleman's books on the subject.

Another — even better — option is a Chinese greenhouse. Chinese greenhouses represent a major leap forward in greenhouse design. These greenhouses are made for growing vegetables — not just cold weather veggies like lettuce — in cold climates all winter long without outside heat.

Chinese greenhouses are earth-sheltered on the north, east, and west sides. They are covered at night to keep heat in. (You can find a little information on Chinese greenhouses online.)

Another option for year-round food production is aquaponics. Aquaponics systems combine fish and vegetables. Vegetables are grown either in small pots on floating rafts or in pea-gravel beds; they are nourished by "waste" water from fish living in nearby tanks.

Aquaponics systems allow homeowners to produce animal protein and a wide range of vegetables. Systems can be set up indoors or in Chinese greenhouses for year-round production.

Month Eleven: Achieving Health Self-sufficiency

The topics I've covered so far focus on what most survivalists and folks interested in self-sufficiency see as the nuts and bolts of personal independence. If you want to achieve total or near-complete self-sufficiency, however, you may need to rethink some other key issues, notably your health and fitness. Let's start with health and eating habits that lead to bad health.

Eating a healthy diet along with daily exercise can help you lose weight and improve your health — and, they will help you reduce your dependence on expensive medicine, costly doctors, and outrageously expensive hospitals. Obesity and poor diet and lack of exercise lead to high blood pressure, high cholesterol levels, hardening of the arteries, stroke, heart attacks, and a variety of cancers — all requiring expensive medical intervention. They can also lead to arthritis in key weight-bearing joints such as the knees and hips that may require expensive artificial joint replacement. Excess weight leads to hernias, varicose veins, and acid reflux — all requiring costly medical treatment. What is more, excess pounds make us more sedentary, which creates a vicious negative downward spiral that increases weight gain and all the associated medical problems.

Eating right is a tall order and won't be achieved in a month. But you can start now. (That's why I suggested you begin implementing ideas for healthy living as you take care of the nuts and bolts of sustainability like energy, food, and water.)

If you truly want to be self-sufficient with respect to your health, it's important to change your diet. Eat less and eat a much more healthy diet.

Here's the rundown on what you need to do to lose weight and improve your health: (a) begin eating whole grain foods (whole wheat bread and pasta and grains like brown rice); (b) slowly but surely reduce the amount of animal protein you consume, especially red meat, unless it's from grass-fed cattle; (c) begin obtaining protein you need from plant-based foods like nuts, seeds, and tofu; (d) dramatically increase your consumption of onions, mushrooms, beans, greens, and berries as they contain antioxidants and phytochemicals that will virtually eliminate your chances of contracting cancer, cardiovascular disease, late-onset diabetes, high blood pressure, etc. Check out Dr. Joel Fuhrman's excellent books and DVDs. As a long-time student of nutrition who received a Ph.D. from the University of Kansas School of Medicine and an author of a college-level human biology textbook, I can say with authority that Dr. Fuhrman's recommendations is the soundest dietary advice on the planet. It's based on sound science, not dietary nonsense, like the vast majority of diet books.

Eating right will take many years, so start slowly by weaning yourself off of junk food — cake, candy, ice cream, burgers, and fries — and then eating the foods that will provide nutrients you need without all the extra fat-generating heart-clogging calories.

Month Twelve: Get Active

As you are learning to eat right, it's a good idea to direct your attention to your physical fitness. Here's my advice: get off your butt and get active. Get in better shape; but if you are obese or overweight, start slowly. Do as many chores as you can by hand — without costly, energy-consuming equipment — to stay in shape and lower your chances of disease. Walk or ride a bike to the store, don't drive. Rake leaves, don't blow them. If you are living in the country and are raising livestock, leave the ATV behind and walk your fences to check for damage. Carry buckets of feed and bags of grain to your animals, don't truck them there. Here's the main idea to live by: substitute muscle power for gas power. But, please, work in conjunction with your family doctor.

Month Thirteen: Economic Self-sufficiency and Interdependence

You can improve your health and reduce your dependence on our "modern" health care system, which is designed primarily to treat, not prevent, disease. You can generate your own energy to power your home, catch rainwater, grow your own food, and can and freeze food, but you still may need resources and goods others have to offer.

To ensure you can meet all of your needs, begin to develop a network of like-minded friends and associates that you can trade with, should times get tough. Bartering could become a valuable currency.

Remember, you can trade excess firewood, fruit, and vegetables for help building fences or trimming trees. Or, your excess can be used to barter for tools you need to borrow. Skills in health care, preventive medicine, carpentry, plumbing, electrical wiring, solar installation, small equipment repair, auto repair, bike repair, energy efficiency, gardening, and composting are valuable currency with which you can barter among your circle of friends. My advice: develop as many skills as you can so you can be more self-sufficient and have a valuable commodity with which to barter. Encourage friends to develop a wide range of skills, too. It's not as difficult as you may think.

You can learn skills by taking workshops, reading, or helping friends work on various projects. In the end, a network of like-minded individuals who are all striving for self-sufficiency can be an extremely valuable asset. You very likely won't survive alone in a disaster. You'll need others.

Month Fourteen: Sustainable Transportation

My last recommendation is to develop several forms of sustainable transportation — ways of moving about and moving goods that require little, if any, fuel or use a type of fuel that is relatively easy to acquire. A horse might suffice, but they do require an awful lot of food.

Bicycles are an even better option. They'll help you improve and maintain your health and require very little energy — just a granola bar or two. A bicycle equipped with a large basket will enable you to deliver surplus food to friends and carry home the tools or canned goods for which you've traded your vegetables.

Another option to consider is an electric car. Electric cars on the market today are ideal for 50 to 100 mile jaunts. Chances are, that's all you really need. Currently, you can get some awesome deals on used electric cars — they're often selling for a third of the price of a new one!

Electric cars can be charged by a solar array or a wind system, so you won't have to purchase fuel at $10.00 per gallon when the shit hits the fan. What is more, electric motors are reliable and very low maintenance. No oil to change or fluids to add.

Conclusion

At this point, the ball's in your court. It's time to start designing and building a composting toilet and greywater system so you can take a bold step on the path toward a sustainable, self-sufficient lifestyle. I wish you well and look forward to meeting many of you at upcoming events and perhaps at our educational center. Many readers have shared their success stories with me, and I'm always thankful when people reach out to me to share ideas, challenges they've faced, photos of successful projects, and success stories. Take care, and be well. You can reach me through danchiras@evergreeninstitute.org.

Index

Page numbers in *italics* indicate illustrations and photographs.

H

About the Author

Dan Chiras is a highly respected educator and the internationally acclaimed author of over 30 books on residential renewable energy, green building, and sustainability, including *The Homeowner's Guide to Renewable Energy* and *Power from the Sun*. He has studied and written extensively about ecologically sound wastewater systems. Dan is the president of Sustainable Systems Design and the director and lead instructor at The Evergreen Institute's Center for Renewable Energy and Green Building (www.evergreeninstitute.org), where he teaches workshops on sustainable waste management, self-sufficiency, energy efficiency, solar electric and solar thermal systems, wind energy, green building, natural plasters, and natural building.